Belt conveying of minerals

Related titles:

Handbook of gold exploration and evaluation
(ISBN 978-1-84569-175-2)
This important book covers the nature and history of gold, geology of gold ore deposits, gold deposition in the weathering environment, sedimentation and detrital gold, gold exploration, lateritic and placer gold sampling, mine planning and practice, metallurgical processes and design, and evaluation, risk and feasibility. The breadth of subject matter makes this a standard reference work on alluvial gold deposits, exploration and mining.

Mining in the Americas: Stories and history
(ISBN 978-1-85573-131-8)
Twenty years of work went into the writing of this: the first book to cover the history of mines and mining in North and South America. The text is enlivened by sketches of many miners the author got to know over the decades.

Valuing mining companies: A guide to the assessment and evaluation of assets, performance and prospects
(ISBN 978-1-85573-435-7)
Valuing mining companies gives a perspective on the international mining and metals industry, from historical details of mines and mining to the potential problems encountered in attempting to value a mining company. Chapters are devoted to accounting practices and taxation, providing valuable information on the financial aspects of mining. It also covers various techniques used to value mining companies and gives instruction on how to set up a portfolio and begin trading in this complex field.

Details of these and other Woodhead Publishing books, as well as books from Maney Publishing, can be obtained by:

- visiting our web site at www.woodheadpublishing.com
- contacting Customer Services (e-mail: sales@woodhead-publishing.com; fax: +44 (0) 1223 893694; tel.: +44 (0) 1223 891358 ext. 130; address: Woodhead Publishing Limited, Abington Hall, Abington, Cambridge CB21 6AH, England)

If you would like to receive information on forthcoming titles, please send your address details to: Francis Dodds (address, tel. and fax as above; e-mail: francisd@woodhead-publishing.com). Please confirm which subject areas you are interested in.

Maney currently publishes 16 peer-reviewed materials science and engineering journals. For further information visit www.maney.co.uk/journals.

Belt conveying of minerals

E. D. Yardley and L. R. Stace

Woodhead Publishing and Maney Publishing
on behalf of
The Institute of Materials, Minerals & Mining

CRC Press
Boca Raton Boston New York Washington, DC

WOODHEAD PUBLISHING LIMITED
Cambridge England

Woodhead Publishing Limited and Maney Publishing Limited on behalf of
The Institute of Materials, Minerals & Mining

Published by Woodhead Publishing Limited, Abington Hall, Abington,
Cambridge CB21 6AH, England
www.woodheadpublishing.com

Published in North America by CRC Press LLC, 6000 Broken Sound Parkway, NW,
Suite 300, Boca Raton, FL 33487, USA

First published 2008, Woodhead Publishing Limited and CRC Press LLC
© 2008, Woodhead Publishing Limited
The authors have asserted their moral rights.

This book contains information obtained from authentic and highly regarded sources.
Reprinted material is quoted with permission, and sources are indicated. Reasonable
efforts have been made to publish reliable data and information, but the authors and
the publishers cannot assume responsibility for the validity of all materials. Neither the
authors nor the publishers, nor anyone else associated with this publication, shall be
liable for any loss, damage or liability directly or indirectly caused or alleged to be
caused by this book.
   Neither this book nor any part may be reproduced or transmitted in any form or by
any means, electronic or mechanical, including photocopying, microfilming and
recording, or by any information storage or retrieval system, without permission in
writing from Woodhead Publishing Limited.
   The consent of Woodhead Publishing Limited does not extend to copying for general
distribution, for promotion, for creating new works, or for resale. Specific permission
must be obtained in writing from Woodhead Publishing Limited for such copying.

Trademark notice: Product or corporate names may be trademarks or registered
trademarks, and are used only for identification and explanation, without intent to
infringe.

British Library Cataloguing in Publication Data
A catalogue record for this book is available from the British Library.

Library of Congress Cataloging in Publication Data
A catalog record for this book is available from the Library of Congress.

Woodhead Publishing Limited ISBN 978-1-84569-230-8 (book)
Woodhead Publishing Limited ISBN 978-1-84569-430-2 (e-book)
CRC Press ISBN 978-1-4200-7606-6
CRC Press order number WP7606

The publishers' policy is to use permanent paper from mills that operate a sustainable
forestry policy, and which has been manufactured from pulp which is processed using
acid-free and elementary chlorine-free practices. Furthermore, the publishers ensure
that the text paper and cover board used have met acceptable environmental
accreditation standards.

Project managed by Macfarlane Production Services, Dunstable, Bedfordshire, England
(macfarl@aol.com)
Typeset by Godiva Publishing Services Limited, Coventry, West Midlands, England
Printed by TJ International Limited, Padstow, Cornwall, England

For the British coal mining industry

Contents

1	Introduction	1
2	History and economics of conveyor applications	4
2.1	Early applications of belt conveyors	4
2.2	Belt conveyors in mines	5
2.3	Belt conveyors in stone quarries and other surface mines	9
2.4	References	15
3	Design of belt conveyors 1 – power requirements and belt tensions	17
3.1	Introduction	17
3.2	Some basic considerations	19
3.3	Belt capacity	21
3.4	Power requirements	21
3.5	Belt tensions	29
3.6	Tension changes over the belt width	39
3.7	Concluding remarks	42
3.8	References	42
4	Design of belt conveyors 2 – hardware (idlers, structure, pulleys, drives, tensioning devices, transfer points and belt cleaning)	44
4.1	Introduction	44
4.2	Conveyor idlers	44
4.3	Conveyor structure	51
4.4	Pulleys	52
4.5	Conveyor drives	54
4.6	Methods of tensioning belts	60
4.7	Design of transfer points	62

Contents

4.8	Belt cleaning	66
4.9	High angle conveyors	68
4.10	References	69

5 Belt constructions — 71

5.1	Introduction	71
5.2	Textile carcase belts	71
5.3	Steel cord belts	78
5.4	Cable belts	82
5.5	References	83

6 Joining conveyor belts — 84

6.1	Introduction	84
6.2	Mechanical fasteners	84
6.3	Spliced joints	89
6.4	Concluding remarks	93
6.5	References	94

7 Standards, test methods and their standardisation — 95

7.1	Introduction	95
7.2	General remarks	95
7.3	The standardisation process	96
7.4	Specific standards and tests standardised by the International Organisation for Standardisation (ISO) and the European Committee for Standardisation (CEN)	98
7.5	Other tests	105
7.6	Concluding remarks	107
7.7	References	108

8 Safety considerations 1 – fire and electrical resistance properties of the belt conveyor — 109

8.1	Introduction	109
8.2	Fire hazards	109
8.3	The Cresswell disaster	111
8.4	Early research into conveyor fires	112
8.5	Questions regarding Barclay's approach	119
8.6	The European dimension	121
8.7	Safe enough?	125
8.8	Australian studies	128
8.9	Mid-scale galleries	133
8.10	Concluding remarks on conveyor fire safety	135

8.11	Electrostatic hazards	137
8.12	References	140

9 Safety considerations 2 – nip points, stored tension, man-riding and materials transportation on belts   142

9.1	Introduction	142
9.2	Nip point accidents	142
9.3	Stored energy	146
9.4	Man-riding	146
9.5	Materials handling by belt conveyor	150
9.6	General comments	150
9.7	References	151

10 Maintenance and monitoring   152

10.1	Introduction	152
10.2	Supply, storage and handling of belts	152
10.3	Belt tracking or training	153
10.4	Optimising belt life	154
10.5	Monitoring the condition of belts	155
10.6	References	157

11 Case histories   158

11.1	Selby mine	158
11.2	Prosper-Haniel	161
11.3	ATH Resources	162
11.4	'The biggest and the best'	165
11.5	References	167

Appendix 1   Derivation of belt capacity   168

Appendix 2   Listing of international, European and national standards relating to belt conveyors   171

Appendix 3   Man-riding conveyors – précis of conditions set out in British Coal Corporation 'Codes and Rules Underground belt conveyors' CR/13   180

A3.1	Introduction	180
A3.2	Clearances	180
A3.3	Key dimensions	181
A3.4	Men and minerals	181
A3.5	Inspection roles	181

A3.6	Boarding and alighting stations	181
A3.7	Safety devices	182

Index   **183**

1
Introduction

Belt conveyors used to transport minerals are to be found all around the world in a large number of surface and underground mining operations. The idea of using the conveyor belt is not new, indeed, the first belt conveyors were introduced at the end of the nineteenth century; the basic principles of operation have not changed. However, over the years the capacity rating of belt systems and the length over which material can be transported have increased very considerably, together with the power inputs, the size of components and the degree of sophistication.

As with any mature technology, there are many publications on belt conveyor systems. Some are written by trade associations and equipment manufacturers, which, naturally, tend to favour their own products and systems. Others, which are of wider scope than this book, do not deal with certain aspects that we considered valuable. Yet others are of limited availability. There are also considerable numbers of papers in the technical press, but these tend to be written by experts for experts. The authors of this book, therefore, have been aiming to produce a publication that gives a balanced view on the technical issues associated with belt conveyors, adding something new where there is something new to say. Although we have had to cover some of the basic ground, we have tried to do this in a way that does not merely repeat what already appears in some of the excellent publications that exist. The book is intended to be both useful and accessible to mine and quarry operators that use or are considering using belt conveyor systems, and also to students in those engineering disciplines that will ultimately lead to employment in the mining and quarrying industry.

There are three key motivators within this general aim.

The first of these is the need, as we see it, to preserve information that is in danger of becoming forgotten or lost and which is required to prevent operators spending too much time 're-inventing the wheel'. It will become fairly obvious to the reader that both authors, one from a mining engineering background and the other, a mechanical engineer, spent most of their working lives in the UK deep coal mining industry during a period when the industry was of significant size and when the drivers for excellence in technical development were strong. Following the privatisation of the UK coal mining industry in 1995, the number of operating conveyors in the industry has reduced considerably, as the number of working mines has reduced to a handful. Whilst there remain experienced and skilled engineers working in this industry, their numbers are diminishing and a generation of engineers that oversaw the technical innovations made during the period from 1960 to 1980 has retired. Every piece of good practice that has been adopted today has been adopted because of the accumulated knowledge and experience of many years of operations with conveyors. The authors recognised the danger of this experience being lost and wished to explain in this book as often as possible, 'why things are the way they are'.

Belt conveying equipment is highly standardised. That is, it is the subject of a seemingly bewildering plethora of standards, whether International, European or National. We have attempted to give a reference guide to those features that have been standardised and where that information can be found. We have also sought to identify when standards give conflicting information.

The final objective is to highlight the need for best practice in conveyor design and operation such that high levels of operational safety are maintained. Conveyor belt systems can represent a significant operational hazard. Studies of accidents associated with belt conveying demonstrate how the causes of injury and incident are concentrated into a few types. This repetitive nature of accidents is worrying to mine and quarry operators and to regulatory authorities alike and exposure of this phenomenon has been an important aim of this book.

The content of the book commences with a look of the history of belt conveying of minerals and a brief review of the technical and economic issues behind two competitions. The first was that between the conveyor belt and rail-borne transport of coal in deep mines. This took place in the post-war years. The second is currently taking place in hard rock deep mines and surface mines, including the stone quarries of the UK, between conveyor belts and trucks. The book continues by reviewing and comparing the standardised methods of designing capacities and power requirements of belt conveyors and the management of belt tension. The fourth chapter continues the design theme by considering issues associated with the hardware from which the conveyor system is constructed. Chapters 5 and 6 look at the manufacture and design of the belting itself and the methods available to join lengths of belting effectively. In Chapter 7, a review is made of the standardised mechanical test methods applied around the world for belting to ensure fitness for purpose.

Chapters 8 and 9 look at the operational safety of conveyor systems. In Chapter 8 particular note is made of the need, in many circumstances, for the fire-resistant properties of conveyor belting giving the historical context of the development of test standards for fire-resistant conveyor belting in Europe and elsewhere in the world. In Chapter 9 the problems associated with the interaction between personnel and belt conveyor systems, particularly those associated with physical injury, are highlighted.

In Chapter 10, essential maintenance and monitoring of belt systems to ensure safe and efficient operations is considered, whilst the book concludes at Chapter 11 with a more detailed review of a number of significant case studies of belt conveyor applications chosen by the authors.

It is our hope that, after having read this book, the mine or quarry engineer will feel more confident to sit down with designers, manufacturers and suppliers of belt conveying equipment and make informed judgements about competing conveying products and strategies.

The authors wish to acknowledge the assistance they have received in writing this book from a number of companies, including Fenner Dunlop, Continental Conveyors, Flexco and ATH Resources Ltd. We also offer special thanks to Brian Rothesy, Alan Kirk and Mansell Williams for their significant personal contributions during our research for and the writing of this book.

2
History and economics of conveyor applications

> The introduction of conveyors in 1902 represented an important step in mining, as their use ... lightens considerably the work of the miner.
>
> I. C. F. Statham, *Coal-Mining*

2.1 Early applications of belt conveyors

Historically, the first conveyor belts were made from layers, or plies, of canvas or woven cotton fabric, termed cotton duck, with natural rubber interlayers (interplies) and natural rubber covers. According to Streets[2.1] the first documented installations of belt conveyors were in granaries built by the Mersey Docks and Harbour Board in Liverpool, UK in 1868. The belts used on these installations consisted of two plies of canvas faced by natural rubber. These belts had short lives, not because of their construction but because of their use with rollers that had flared ends to form a trough. This resulted in the belt contacting the rollers at different linear speeds and rapidly wearing out. Flat rollers were then used, giving greater belt life but lower conveyor capacity. Streets reports that, because of the high price of rubber, it became impossible to produce rubber faced canvas belting economically and lower cost 'solid woven' cotton belts, i.e., without plies or interlayers, were used for many years. These belts were said to have the disadvantages of being prone to damage, the absorption of moisture and being less flexible than the ply belts previously used.

Fayed and Skocir[2.2] report that the first true modern conveyor having troughed idlers dates back to 1891 in Ogden iron ore mine in New Jersey. They state that Thomas Edison assisted in the design of the troughed idler. However, Streets states that Robbins invented the troughing idler in 1885.

Following the introduction of conveyors into coal mines in 1905 by Richard Sutcliffe Ltd and the greater availability of rubber after the 1914–18 war, rubber faced cotton ply belts became the norm in belt conveyors. The use of cotton fabric had resulted in limitations to the load carrying capacity and length of conveyors that could be installed because of the limitation in the strength of belts that could be produced. On long runs it had been the practice to install a series of conveyors, each with its own drive, associated equipment and transfer and loading installations. Improvements in the strength and performance of rubber and canvas multiple plied belts enabled longer installations to be planned and more powerful drives to be installed with the elimination of the multiplicity of drives and other equipment, and with consequent savings in capital installation costs. Streets[2.1] reports that the first conveyor using rubber and canvas ply belting to exceed a mile (1.6 km) in length was installed in 1944. To give some idea of the duties of conveyors at the time that Streets was writing his book in the mid-1950s, cotton fabric, plied belts with typically seven plies (but up to 15), were being used with drives of around 120 horsepower (90 kW) on conveyors up to one mile (1.6 km) in length. He claims that the longest conveyor in the world at that time using rubber and canvas ply belting was 6540 feet (1.24 miles, 1.98 km) in length. At this time conveyor belts were specified according to the weight of the cotton fabric (cotton duck), used in their manufacture and the number of plies used. The weight used was that of a piece of fabric 36 inches (914 mm) wide and 42 inches (1067 mm) long. The most common weight of duck used was 32 ounces (907 grammes), but 28 ounce (794 g), 36 ounce (1021 g), 42 ounce (1191 g) and 48 ounce (1361 g) ducks were also used. The introduction of synthetic fibres into the weaving of belt carcases during the 1960s led to a considerable increase in belt strengths and allowed the use of much longer conveyors. The use of duck weights was discontinued and the specification of belts became based on strength levels rather than fabric weights.

2.2   Belt conveyors in mines

Professor I C F Statham[2.3] in his book *Coal-Mining*, published in 1951, stated that

> The introduction of conveyors in 1902 represented an important step in mining, as their use eliminates the necessity for filling coal direct into tubs and therefore lightens considerably the work of the miner.

He wrote that the advantages of conveying, when compared to the use of rail mounted vehicles for minerals transport, which was the most widely used alternative at the time, included increased productivity, a reduction in the number of roadways driven, no rail laying or tub movements by hand, better distribution of loading points along the face and more systematic systems of work generally. Against these advantages were set the increased capital and running costs associated with the use of conveyors.

2.2.1   Belt conveyors at the coalface

By 1948, belt conveyors were being applied on 5428 of 8515 coalfaces in UK mines. Coal was normally loaded onto the bottom belt at floor level. A simple

6   Belt conveying of minerals

2.1 Coalface belt conveyor (former NCB photograph – published by kind permission of Department for Business, Enterprise and Regulatory Reform).

plough arrangement allowed the coal to be transferred at right-angles onto the gate conveyor[1] at the main gate[2] end of the face. Statham reported that such conveyors usually only extended for 100 yards (91 m), but could exceptionally be of a length up to 440 yards (402 m). Belts were relatively narrow by modern standards and ran at slow speeds. He stated that a 51 cm wide belt running at 0.5 m/s would transport 49 tonnes/hour. This statistic alone, when compared to today's projected capacities measuring thousands of tonnes per hour, emphasises how conveyor belt technology has changed since then.

Coalface belt conveyors were frequently not fitted with intermediate rollers between the two ends, termed the head and tail ends. (The head end was the point of delivery of the coalface conveyor onto the gate conveyor and the tail end was the point at which the belt ran around the return roller.) Especially in thinner seams, the bottom belt ran in contact with the floor and the return strand was supported by steel cross members slung between roof supports near the roof of the seam (Fig. 2.1). Belt breakages were commonplace because frequently the oldest conveyor belt was re-used on the coalface. It was also common for the belt to become overloaded by large lumps of coal. The temptation for miners to load the belt with large lumps of coal, rather than break them up before loading, a hard manual task, often resulted in the lump sliding off the conveyor into other people's working areas nearer to the main gate, the trapping of the lumps along the face on supports, belt structure or at the delivery plough, or the lumps falling off the gate road conveyor. As the mining sequence was cyclic, the whole conveyor was dismantled every 24 hours and rebuilt in a new track as the face advanced.

1. Gate conveyor – a belt conveyor running in a roadway or tunnel directly leading to a coalface.
2. Main gate – the roadway or tunnel in which the belt conveyor directly serving the coalface ran.

History and economics of conveyor applications 7

Mechanisation of all coalface operations during the 1950s and 1960s swept away these applications at the coalface. But the conveyor had arrived and found more significant applications elsewhere in the mine. Statham[2.3] noted over 3500 gate belts installed in UK coal mines. Gate belts at this time were between 50 cm and 75 cm in width and ran at speeds of 0.75 m/s to 1 m/s. He stated with some prescience, 'There is no doubt that, in the future, transport of coal from the face will be largely by belt conveyor'. How right he has been proved to be!

2.2.2 Mine trunk conveyors[3]

Statham described a similar story with trunk conveyor belts. Some 1000 were working in UK mines in 1947, but single runs were limited to some 1400 m between head and tail end in level roadways. Belt speeds of up to 2 m/s with wider belts increased capacities up to 185 tonne/hr. Despite the great success of the belt conveyor and its proclaimed advantages, one recognised disadvantage was that mines still required additional transport systems to bring materials, men and equipment into the mine. He did note, however, that the belt could be used to transport men. More can be found about this topic in Chapter 9.

One factor that now came into play to affect the spread in applications of the belt conveyor in the UK coal mining industry was the major re-construction, modernisation and new mine development programme of the 1950s. Many of these mines were designed on the Continental European principle of Horizon mining. This development had been singled out in the Reid Report[2.4] which had highlighted the use of locomotive haulage on straight level roads, instead of traditional rope haulage systems. In Horizon mining, major developments radiated from the shaft as 'horizons' on level gradients to intersect the seams near to the working areas.[2.5] This was ideal for locomotive haulage of mineral out of the mine and materials and men into the mine, replacing the older rope haulage. However, the major coal reserves in the UK, in the North East, South Wales, and particularly in Yorkshire and the East Midlands, were blessed with only moderate strata gradients that better suited development of the mine following seam level.

In the 1960s and 1970s, the idea of the surface drift mine began to dominate the development of new mines and the re-development of existing mines. This was an idea that had already been put into practice in Yorkshire as early as 1927[2.6] but that was to find its full expression under the government's Plan for Coal in the 1970s. Instead of both entering the mine and removing the mineral through vertical shafts, mine access and coal clearance could be more efficiently achieved by the use of inclined tunnels (or drifts) driven at gradients appropriate for the application of systems of conveyor belts, directly from the mine workings to the surface preparation facility. This was the system of mine design

---

3. Trunk conveyor – a belt conveyor running in a major mine road or tunnel, often fed by mineral from one or more gate conveyors.

found most commonly in the United States and Australia and the replacement of intermittent shaft haulage of coal from the mine to a continuous conveyor system was a major driver to improve and uprate belt conveyor design and operational efficiency. A commentator[2.6] writing in 1987 stated:

> with the introduction of a modern breed of conveyor belting, the effective winding capacity through a drift can exceed 2000 tph – nearly three times greater than that achievable conventionally at a normal shaft.

As with nearly every aspect of coal mining, although the fundamentals of the longwall[4] coalface and its coal clearance system has changed little, the production expectations from each face and the size and capabilities of the equipment used has increased almost exponentially.

There were a number of features of belt haulage of minerals that concerned mine operators. Despite their theoretical ability to clear minerals continuously from the mine, when the conveyor was halted, for any reason, coalface operations came to a halt. It was recognised that much production time was being lost, primarily because the 'belts never ran', i.e., there were frequent long and short delays due to outbye[5] belts being stopped. With locomotive systems, the mine cars themselves had acted as a buffer between the production process and the clearance system, allowing much more continuous production. With conveyor belts, outbye delays had a much more direct and costly effect on coal output levels. The prime causes of delay were blocked chutes at transfer points and belt damage due to the passage of large lumps of mineral through the system.

Belt transport systems were also heavy on their demands for manpower. In the 1960s and early 1970s every transfer point required manning, as well as the need for belt patrol men and spillage cleaners. Developing technology overcame these problems.

It is notable that there were also problems in matching the capacity of the coal clearance system with the peak producing capacity of the coalface. This was simply because mine operators focused their attention (and investment) on higher capacity coalface equipment and often ignored, or were unable to deal with, the infrastructure backing up the coalface. A reduced number of higher productivity coalfaces led to less complex belt systems with fewer transfer points. Conveying technology developed to increase belt capacity by faster conveying speeds, more powerful drives, stronger and wider belting and longer single runs. Advances in sensor and control technology removed the source of many minor stoppages and allowed remote operation of belts from central control rooms. These developments removed the need for the man at the transfer and productivity was increased. Coal sizers were placed strategically to reduce the 'large lump' problem. Balancing of production to coal clearance capacity

---

4. Longwall – a system of mining coal.
5. Outbye – in a direction away from the coalface towards the pit bottom.

was provided by mechanical and strata bunkers that were introduced into the transport system to smooth out peaks and troughs in the flow of mineral, and also to allow separation between flows of coal and development stone[6] prior to movement out of the mine.

In the 1960s major trunk and surface drift belts were installed to handle the output from a number of, often widely spread, faces. By the year 2000, individual longwall faces were designed to produce 4 million tonnes per annum and all of this output ran along one belt system. Gate roadways were getting longer – therefore gate belts were designed for greater lengths (4 km or longer) and very high output levels, well in keeping with outputs conveyed along fixed length main trunk conveyors. For example, one of the newer Australian mines, Dendrobium in New South Wales, has been equipped with gate conveyors rated at 4500 tonnes per hour and trunk conveyors rated at 6000 tonnes per hour.[2.7]

Watt in 1979[2.8] referred to a German study that stated that for a daily output of 10 000 tonnes, locomotive haulage of coal would have a capital advantage over belt conveyors for distances greater than 4 km and an operating cost advantage for distances greater than 2.5 km. Despite studies such as these, in practice, haulage of coal by conveyor won the battle of conveyor belt versus locomotive systems and in major coal mining applications around the world today, it is unusual to find mines transporting coal by locomotive systems.

2.3 Belt conveyors in stone quarries and other surface mines

The battle of belt conveyor versus other forms of transport, notably truck haulage, is still being fought in other mining and quarrying applications.

Hartmann[2.9] compared the advantages and disadvantages of principal haulage systems available over a range of mining applications wider than just underground coal mining. He noted that conveyors had the advantages of high, continuous output, ability to operate over a range of grades and, perhaps most significantly, low operating costs. Their disadvantages are that conveyor systems are seen as inflexible, high cost in terms of initial investment and having limitations relating to the size of material that can be transported efficiently. This final disadvantage is the reason that belt conveying haulage has been slower to be applied in both underground and surface mines where the prime means of production is breaking rock using explosives.

Hartmann published comparative cost data for the different types of haulage in a surface iron-ore mine with an 8.9 km haul. Overall costs of rail haulage were some 85% of those of truck haulage, but belt conveyor haulage costs were even lower at 65% of those of trucks.

---

6. Development stone – rock mined when forming roadways or tunnels.

The 'Rules of Thumb' set out by McIntosh Engineering[2.10] underline the advantage of belt conveying across the spectrum of operations. They state:

- An underground mine is more economically served by a belt conveyor than railcars or trucks when the daily mine production exceeds 5000 tons.
- As a rule, belt conveyor operation is more economical than truck haulage if the conveying distance exceeds 1 km.

A further attempt to provide guidelines for the selection of forms of haulage on cost grounds can be found in the Conveyor Equipment Manufacturers Association (CEMA) publication.[2.11] CEMA agrees with the second bullet point above and adds four other economic points.

- Beyond 1 km distance, the ton-mile cost of transport by belt conveyor may be as low as $1/10^{th}$ cost by haul truck.
- Estimated operating maintenance cost per year for a belt conveyor is 2% of the purchase cost plus 5% of the belt cost.
- Belt replacement is on average every 5 years for hard rock applications and up to 15 years for non-abrasive applications.
- Well maintained conveyor systems can reliably operate at 90% or higher availability.

The third of the bullet points above also underlines the factor that belt conveyors are more likely to be applied to convey relatively low abrasive minerals and also minerals that have been reduced by crushing or grading to handleable sizes. The difficulty of conveying mixed size product with occasional large lumps forms one of the biggest restrictions to the application of belt conveyor transport in a wide range of mineral extraction sites. In underground coal mines, the longwall mining process in which coal is cut rather than blasted using explosives, forms the first part of product preparation. Shearer power loaders[7] routinely are fitted with sizers to create a product that can be handled easily in transport. This is usually augmented by a second breaker unit housed in the stage loader[8] before the product is loaded onto the belt conveyor system. In hard rock mines and stone quarries, there is little possibility of hauling by conveyor unless such preliminary sizing takes place. Therefore, in the typical UK hard rock quarry employing blasting as the rock winning technique, and out-of-pit primary stage crushing, off highway trucks remain the preferred method of transport. With payloads of around 70–100 tonnes, together with a degree of operational flexibility far in excess of fixed position belt conveyors, trucks still predominate.

Material size distribution is not the only concern when transporting mineral by belt conveyor. It is also necessary to load material at a consistent rate on to the belt, primarily for efficiency of transport and also to restrict spillage.

7. Shearer power loader – a machine used to cut and load coal on a longwall coalface.
8. Stage loader – chain scraper conveyor delivering coal from the coalface onto the gate conveyor.

A number of authors have reported the spread of conveyors in quarries increasing. A recent article stated: 'When there is a requirement to transfer large volumes of bulk material over a fixed distance, whether less than 200 m or greater than 20 km, belt conveyor systems have proved themselves to be significantly more cost efficient and environmentally acceptable than truck haulage.'[2.12]

Haulage of mineral by trucks does carry some significant costs. Trueman[2.13] noted the following points:

- Only some 40% of the energy consumed in truck operations is expended hauling payload. The remainder is employed hauling the truck body itself.
- Trucks tend to be empty on the return journey.
- Conveyors consume some 80% of their energy delivering payload.
- Energy costs for trucks are 3 times greater than for conveyors on the level and up to 8 times greater lifting the payload out of the pit.

Whilst the ability of a conveying system to operate continuously is a very significant factor in the decision-making process, in surface operations where environmental concerns are never far from an operator's mind, belts can be housed in continuous enclosures retaining dust and noise within and ensuring a dry product. The environmental footprint of a truck operation when it comes into the proximity of the general public is quite significant.

In an effort to reduce the haulage costs, quarry operators have increased truck size to reduce the number of journeys (cycles) and sought to reduce the distance for blasted rock to be transported between face and primary crushing. In this option, crushed rather than blasted rock can then be conveyed by belt from the quarry. This has resulted in the construction of major fixed in-pit crushers within the quarry itself (Fig. 2.2).

2.2 Primary crusher and belt conveyor feed to the screening plant in a granite quarry.

An example of belt conveyors finding a successful application in a hard rock quarry can be found at Midland Quarry Products Cliffe Hill granite quarries in Leicestershire, UK. When operations relocated from the Old Cliffe Hill quarry to the New Cliffe Hill quarry in 1980s, and a new crushing and screening plant for rock was built, reserves remained at the original site. In order to access these reserves again but continue to utilise the existing plant at New Cliffe Hill, a 713 m long tunnel was driven between the two sites, being completed in 2005. A new semi-mobile primary crusher was installed at the Old Cliffe Hill site and the product conveyed 1.44 km from the crusher through the tunnel to New Cliffe Hill plant at up to 2500 tonne/h.

2.3.1 In-pit crushing and screening

As quarries subsequently developed away from the site of the crusher and to deeper levels in the deposit being mined, the advantages of fixed position in-pit primary crushing became eroded. As early as the mid-1950s, mobile in-pit crushers were being developed. Early designs tended to be large and their mobility, although technically possible, required special measures and equipment that meant that, in practice, they were infrequently moved. The need was therefore identified for reliable, but more manoeuvrable and cost-effective in-pit crushing and conveying equipment.

Whilst companies have developed individual pieces of equipment to solve this problem, the operator has to take into account how the whole crushing, screening and materials handling processes fit together and into the existing mineral winning method and operational layout.

These design requirements have led to a modern breed of track mounted mobile crushers that are in widespread application in the UK aggregates industry. These units would comprise the following features:

- feed hopper
- vibrating Grizzly feeder
- vibrating fines or waste screen
- fines side-discharge conveyor
- crusher unit
- main product discharge conveyor
- tracks or wheels
- power unit.

The in-pit concept is not confined to the crushing of run-of-quarry rock. Secondary and tertiary crushing and screening units can also be incorporated into the process.

An example of a major capital project to re-equip a quarry to produce a different set of products was reported in a recent article.[2.14] It examined capital investment decisions being made at a quarry in the Midlands. The article

History and economics of conveyor applications 13

describes how the 'traditional' method of production and haulage, with four 50-tonne dump trucks hauling the rock up to 2 km from the face to the processing plant had been replaced by conveyor haulage, both fixed and flexible, and mobile in-pit crushers. The reasons for this decision were primarily related to the cost of operations.

The advantages claimed by proponents of in-pit crushing and screening mirror the claimed benefits of operating conveyors over truck haulage. They include:

- lower capital and operating costs
- reduced plant maintenance
- reduced noise and dust pollution
- increased operational safety
- reduced haul road maintenance.

This approach may not be the solution for all operations but it clearly has its place.

2.3.2 Conveyors in surface mining applications

Conveyor transport also plays a major role in some of the larger volume surface mining applications of Europe, most especially the large-scale lignite mines such as those operated by RWE Power AG in Germany and Mini Maritsa Iztok (MMI) in Bulgaria, where pits measure kilometres across and huge tonnages of lignite are produced. Both companies operate bucket wheel excavators mining the overburden and also the lignite seams. RWE's website claims that the largest bucket wheel excavators can remove 240 000 cu m/d of overburden. They are also claimed to be equipped with the highest capacity conveyor belts currently in use. Belts with a width of 3200 mm and a thickness of 45 mm are fitted and are claimed to have a capacity of 40 000 tonne/h. RWE also claim to operate 266 km of belt conveyors in their four large mines. MMI also have similar dimensioned operations with mining faces up to 10 km in length, all served by conveyor systems. Coal is transported in the pit by conveyor (Fig. 2.3), reaching rail loading outlets outside of the mined area.

The field belt systems of RWE's operations are significant for their widths (2.2–2.8 m) and their speeds (6.5–7.5 m/s).[2.15] Surface mining on this scale relies on the mobility of its belt conveyor systems as well as on their very high capacities. Structure and drives have to be designed to perform at the level of standard fixed length, fixed position systems, but to have the mobility to follow the mining operations (Fig. 2.4).

2.3.3 Conveyors in sand and gravel operations

Sand and gravel working is common in the UK particularly in the Trent Valley area of the Midlands. The workable deposit is normally very close to the surface

14 Belt conveying of minerals

2.3 Surface lignite mine showing conveyors systems.

and around 8 m thick. Extraction is obtained purely by digging out the deposit using shovels or draglines. Workings tend to cover large geographical areas and transport of the extracted product to the plant can form a substantial proportion of the costs of the total operation. Once again operators have to choose between dump trucks and field conveyors to transport minerals and the economic and operational arguments used are very similar to those listed above.

2.4 Mobile conveyor drive for surface lignite mine.

An interesting example of a cost comparison between field conveyors and dump truck operations can be taken from Littler.[2.16] He compared the overall costs of running 600 mm wide and 1000 mm wide field conveyor transport at 200 tonnes per hour and 800 tonnes per hour capacities respectively with articulated dump trucks. In the lower capacity system, the cost savings of conveyor over dump truck was 43%. In the higher capacity system it was 66%.

Littler states that many aspects have to be taken into account when compiling comparisons such as this. He does go on to conclude that field conveyors become more cost effective as the distance from face to plant increases and output rises. He suggests minimum cut offs of 100 tonne/h and 1000 m distances as the lower limit of conveyor efficiency. Similarly, he notes that capital and operating costs of conveyors are such that site lives less than 3 years would also make conveyors uneconomic. It is important to note that with very low value mineral products, such as sand and gravel or crushed rock, savings of pence on costs take on a greater importance.

Littler also notes that flat topography, consistent depth and quality of deposit and rectangular shape of working layout would be required to allow a logical sequence working pattern and planned and regular conveyor moves. If the ratio of sand to gravel, the colour and impurity content of the deposit varies across the reserve, then the flexibility of truck transport can be the key factor in the choice of systems.

The belt conveyor has been a major tool in the development of the mining and quarrying industries. Whilst the fundamentals of the belt conveyor have changed little over the last sixty years of intensive use of this tool in mines and quarries, the engineering of all the components has changed to allow greater carrying capacity, greater mechanical availability, higher levels of operational safety and improved economics. The following chapters of this book will examine in detail some of these developments, why they were made and what their effect has been.

2.4 References

2.1 Streets H *Sutcliffe's Manual of Belt Conveying* W & R Chambers Ltd, 1956.
2.2 Fayed M E and Skocir T S *Mechanical Conveyors – Selection and Operation* Technomic Publishing Company Inc, 1997.
2.3 Statham I C F *Coal-Mining* English Universities Press, 1951.
2.4 Coal Mining Technical Advisory Committee (the Reid Report) Ministry of Fuel and Power, 1945.
2.5 Fritzsche C H and Potts E L J *Horizon Mining* George Allen and Unwin, 1954.
2.6 Standen A W 'Coal clearance in the computer age' *Mining Engineer* 306, March 1987.
2.7 Turner R Presentation on the International Perspective, Seminar on Bulk Handling Conveyors, Midland Institute of Mining Engineers, Sheffield May 2006.

2.8 Watt R G 'Mining transport – short and long term considerations' *Mining Engineer* 211, April 1979.
2.9 Hartmann H L *Introductory Mining Engineering* Wiley, 1987.
2.10 de La Vergne J *Hard Rock Miners Handbook* McIntosh Engineering, June 2000.
2.11 CEMA (Conveying Equipment Manufacturers Association) *Belt Conveyors for Bulk Materials* 6th edition, 2005.
2.12 'Changing conveyors' *Quarry Management*, November 2001.
2.13 Trueman E 'In-Pit Crushing' *Quarry Management*, 25, 7, July 1998.
2.14 'Cloud Hill goes mobile' *Quarry Management*, March 2004.
2.15 Benatsky U 'Topical survey of Rheinbraun AG's Belt Conveyor Systems' *Bulk Materials Handling by Conveyor Belt* III, SME 2000.
2.16 Littler A *Sand and Gravel Production* Institute of Quarrying, 2000.

3
Design of belt conveyors 1 – power requirements and belt tensions

3.1 Introduction

Chapter 2 has shown that belt conveyors can provide an economical means of transporting materials. They can vary in length from a few metres to several kilometres, but all have certain key elements in common. Figure 3.1 shows a typical basic conveyor and defines some of the terms used to describe elements of the conveyor. Essentially an endless belt moves between two points over a series of supporting rollers, propelled by a *driving pulley* or 'drum' and returning via a second pulley. The rollers that carry the belt are called *idlers*. On the upper, carrying strand of the conveyor, where the belt is travelling in the direction of the material being transported, at each support point there are usually three *carrying idlers*. These are arranged in the form of a trough (Fig. 3.2) to increase the carrying capacity of the belt. In general the most common troughing arrangement is for there to be a horizontal *centre idler* and two inclined *wing idlers*. The *angle of trough* or *troughing angle* is defined as the angle between the axis of the wing idler and the horizontal. In terms of conveyor design the material carried is said to have a *surcharge angle* which is the angle between the tangent to the outside edge of the load where it contacts the belt and the horizontal, assuming that in cross-section the load forms an arc of a circle on the belt. On the empty lower, or return, strand there is usually a single *return idler* at each support point. The end of the conveyor where the material is discharged is referred to as the *head end* and the other end, where the material comes onto the conveyor, the *tail* or *return end*. Provision is made for removing slack and applying tension to the belt through a *take-up device*. The conveyor may include *snub pulleys*, which are positioned so as to increase the amount by which the belt wraps around the driving pulley and *bend pulleys* that alter the direction of travel of the belt. The installation may also include a *tripper*, which is a device mounted in either a fixed position or on a travelling carriage for discharge of material from the conveyor at a fixed point or at various points along the conveyor's length. The tripper has a discharge pulley, which is mounted above and in front of a second pulley, the purpose of which is to return the belt to its original course.

18   Belt conveying of minerals

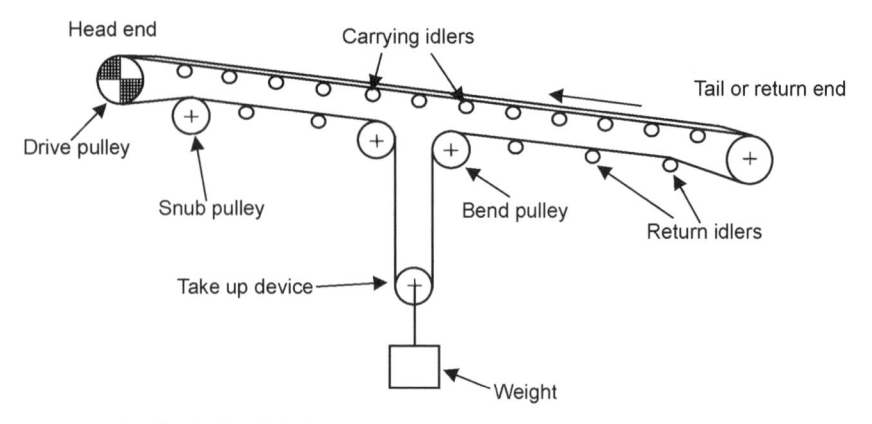

3.1 Typical basic belt conveyor.

One of the most important features of conveyor design is the control and management of the stresses in the belt itself. Since the technology of conveyors is mature and conveyors have been used for many years, standard methods for their design exist. These allow the user to determine all of the parameters needed for his or her design, including calculating belt carrying capacity, belt width and speed, belt tensions and power requirements, and specifying structures, idlers and drive configurations. As will be made clear in the following pages, the design of a conveyor has to be an iterative process, with initial estimates of belt widths, speeds and so on being revised by subsequent calculations. The International Organisation for Standardisation (ISO),[3.1] British Standards Institution (BSI),[3.2] Deutsches Institut für Normung (DIN)[3.3] and the Conveyor Equipment Manufacturers' Association (CEMA) of the USA[3.4] all provide detailed design methods, specifications and analyses. In addition, manufacturers of conveyors and conveyor belts have produced their own comprehensive design methods. Some major conveyor users, such as mining companies, have tended to produce their own standards that refer to the national or international standards but customise them to suit their own particular circumstances.[3.5]

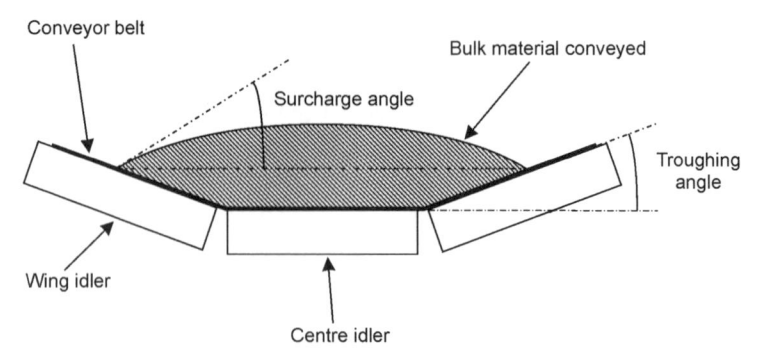

3.2 Cross-section of conveyor belt on a three-idler set.

Design of belt conveyors 1 – power requirements and belt tensions 19

The purpose of this chapter is not to reproduce details of these standards, but to provide the basic concepts involved in the procedures, to compare and contrast the various approaches and to comment on modifications or refinements that might be needed for particular circumstances. The chapter starts with some basic considerations regarding conveyor design and then deals with how belt capacity is determined, calculation of power requirements for conveyors and the derivation of belt tensions. Tension changes over the belt width are then considered, including determination of minimum distances for the transition from a troughed to a flat shape (and vice versa) and the effects of vertical and horizontal curves. The chapter concludes with a discussion on methods of tensioning belts.

3.2 Some basic considerations

The starting point for any conveyor design has to be a knowledge of what is to be moved, how far it is to be moved, the gradient over which it is moved and what constraints there might be to the design of the conveyor. Clearly, for a given troughing angle it is possible to convey the same amount of material on a wide slow belt as on a narrower faster one. The choice of width and speed will be influenced by the nature of the material to be conveyed, available space in the gantry or tunnel and the overall economics of the system. The variety of materials moved by conveyor is vast, ranging from metalliferous ores to limestone, granite, coal, coke, ash, wood pulp and grain. These have an equally extensive range of properties, some of which can affect the design of the conveyor that is intended to handle them. The particular properties that need to be considered are those that might affect the way in which the material feeds onto and off the conveyor, sits on the conveyor and travels on the conveyor. These include size and size range, angularity, abrasiveness and tendency to degrade or to produce dust and whether the material is wet or sticky. Fayed[3.6] notes that CEMA provides materials classification codes that describe materials characteristics for over 500 materials in standard 550:2003 'Classification and Definition of Bulk Materials'.[3.7] This information is also provided in the CEMA handbook,[3.4] which gives detailed explanations and references for the establishment of these codes. As an example of how materials characteristics affect conveyor design, reference 3.5 advises that when choosing belt speeds account needs to be taken of the differential speed of the belt and ventilation air if airborne dust is to be avoided. In UK mines conveying speeds have typically been between 1.5 and 2.5 m/s.

Conveyor belts are supplied in ranges of standardised widths. These ranges depend on in which part of the world the supplier is based. In the USA the standard widths are 18, 24, 30, 38, 42, 48, 54, 60, 72, 84 and 96 inches. ISO 251[3.8] gives eighteen standard widths between 300 and 3200 mm, some of which are clearly metric equivalents of the former inch sizes used in the UK. The choice of belt width may be affected by the size of the lumps of material being conveyed, since the belt needs to contain the material adequately and to avoid material being too close to the belt edge. Streets[3.9] in his book of 1956, advises that in general, the width of the belt is fixed at three times the size of the largest lumps. He goes on to say that if the loading conditions are well designed, it should be possible to load a belt with uniformly sized material that is up to a quarter of the belt width and with mixed material where occasional lumps may be up to half the belt width. If loading is not ideal, these proportions should be changed to one-fifth and one-third respectively. More recent guidance is given in BS 8438[3.2] and by CEMA;[3.5] however, the guidance appears in rather different ways. BS 8438 gives a table of values of maximum lump size for particular belt widths for 'sized' and 'non-sized' material. 'Sized' material is defined as having 95% of the material below the stated size. CEMA provide simple empirical relationships between belt width and lump size for two different surcharge angles and proportions of lumps. For a 20° surcharge angle and '10% lumps/90% fines', the maximum lump size is one-third of the belt width, while for 'all lumps' the maximum lump size is one-fifth of the belt width (compare with Streets' recommendations above). For a 30° surcharge angle the equivalent figures are one-sixth and one-tenth respectively. BS 8438 does not state a surcharge angle, but refers to 'sized' and 'non-sized' material.

Comparisons are difficult because of the different definitions of the conveyed material, but CEMA appears to recommend rather larger minimum belt widths than BS 8438. Table 3.1 shows, for example, the minimum belt widths recommended for a lump size of 400 mm and 20° surcharge angle by BS 8438, CEMA and Streets. CEMA also provides graphs of lump size against belt width for '10% lumps/90% fines' and 'all lumps' at surcharge angles of both 20 and 30°. The belt widths are rather lower for all four circumstances than those derived from the numerical relationships.

Some other factors that need to be considered in the design include the ability of the belt to conform properly to the trough formed by the idlers and the effect on the belt of forming the trough. Reference 3.5, for example, advises that the thicker, stronger belts at the top end of the range of textile carcase belts will only trough and track, i.e., run to a given course properly, at low troughing angles. It also warns that there is a possibility that troughing at 45° may induce excessive stresses in the belt at the transition between the wing and centre idlers and cause

Table 3.1 Comparison of recommended belt widths

	BS 8438	CEMA	Streets 'well-designed'	Streets 'not-ideal'
Non-sized	1000 mm	1200 mm	800 mm	1200 mm
Sized	1400 mm	2000 mm	1600 mm	2000 mm

Note: BS 8438 'non-sized' = CEMA '10% lumps/90% fines' = Streets 'mixed'
BS 8438 'sized' = CEMA 'all lumps' = Streets 'uniformly sized'

Design of belt conveyors 1 – power requirements and belt tensions

premature failure of the belt; indeed the authors are aware of an instance where this occurred and a belt parted along the transitions. Reference 3.5 also advises that higher speeds result in higher wear rates of moving components and the possibility of greater dust make at transfer points.

3.3 Belt capacity

The belt capacity is derived by simple geometry from a diagram such as Fig. 3.2, in which there are three rollers of equal length, although formulae are available for other geometries such as five-idler sets and two-idler sets. All calculation methods assume that the belt is filled uniformly along its length and that the load extends to within a small distance x of the edge of the belt. This distance has been derived empirically and is expressed as a fraction of the belt width plus a constant. Some calculation methods take the length of the centre idler as defining the horizontal, central part of the loaded area, while others, notably CEMA, use an empirical formula based on the central idler length. The ISO, BSI and CEMA methods assume that the material adopts a semi-circular profile (Fig. 3.2). However, DIN 22101 differs from this in assuming a triangular profile and hence a rather larger capacity for a given surcharge angle. Whilst the other methods give virtually identical values for the cross-sectional area, the DIN method gives between 15 and 20% greater area, depending on the troughing and surcharge angles. A detailed calculation of the area of cross-section of the load for a semi-circular profile is given in Appendix 1. Tables giving values of the cross-sectional area for various belt widths, surcharge angles and troughing angles are given in most of the major standards.

Given the cross-sectional area (m²), the load in tonnes per metre run of belt is then given by

$$\frac{\text{cross-sectional area} \times \text{density}}{10^3},$$

when density is measured in kg/m³.

Tonnes per hour is given by:

$$\text{tonnes per metre run} \times \text{belt speed} \times 3600,$$

when belt speed is measured in m/s.

From these relationships an initial choice can then be made of belt width and speed to convey the required amount of material.

3.4 Power requirements

Figure 3.1 shows the basic elements of a conveyor. However, conveyor design can be much more complicated and include loading at various points, changes in slope, downhill sections and multiple drives. All of these and many other factors, such as the design of idlers and structure, belt characteristics and environment can affect the power requirements and belt tensions. The calculation of power and tensions can thus be a very complex process. The bodies referred to in references 3.1 to 3.4 all provide calculation methods, as do conveyor and conveyor belt suppliers. These methods are of varying complexity; indeed, CEMA provide three methods, i.e., Basic, Standard and Universal. Since it is our purpose here only to demonstrate the basic principles, we shall not consider any of these methods in too much detail.

3.4.1 ISO 5048 approach

ISO 5048[3.1] and BS 8438[3.2] are technically identical in their approach to calculating power and tensions, and DIN 22101[3.3] is very similar, differing in relatively minor ways from the other two standards. While ISO 5048 only deals with powers and tensions, the British and German standards have much wider scope and specify the full conveyor design process. ISO 5048 states that it is applicable only to simple conveyors and does not cover complex conveyors that have multiple drives or undulating profiles. BS 8438 states that it is not applicable to underground mining conveyors. All three of the standards start from the premise that the overall resistance to motion of a conveyor is made up of various resistances that can be divided into groups. Typically these may be:

- Main resistances
    - friction in bearings and seals in idlers,
    - due to the movement of the belt from indentation of the belt by idlers, flexing of the belt and movement of the material.
- Secondary resistances
    - inertial and frictional resistances due to the acceleration of the load in the loading area
    - friction on the side walls of chutes in the loading area
    - friction of pulley bearings (not driving pulleys)
    - resistance due to wrapping of the belt round pulleys.
- Special main resistances
    - drag due to forward idler tilt[1]
    - friction from chute flaps or skirts over the full belt length.
- Special secondary resistances
    - friction from belt and pulley cleaners
    - friction from chute flaps or skirts over part belt length
    - resistance due to inverting the return strand of the belt
    - resistance due to discharge ploughs
    - resistance due to trippers.

---

1. Wing idlers may be tilted forward in the direction of belt travel to help belt alignment.

Design of belt conveyors 1 – power requirements and belt tensions

- Slope resistance
  - due to lifting or lowering the load on inclined conveyors.

Of the resistances listed the most important in terms of power consumption are those caused by the idlers, the indentation of the belt by the idlers and movement of the belt. The determination of the resistances is complex and the ISO 5048 approach introduces certain simplifying assumptions in order to enable the power to be calculated. The two assumptions that are made are the introduction of an 'artificial friction coefficient' to allow the evaluation of the main resistances and the introduction of what is essentially a length coefficient to allow the secondary resistances to be calculated. The introduction of the artificial friction coefficient together with the application of Coulomb's friction law, allows quantities that can be measured relatively easily – the masses of the moving items (belt, idlers) – to be substituted for the main resistances which are much more difficult to determine. Then from the equations given in ISO 5048 the main resistance F_H (kN) becomes

$$F_H = f.L.g(q_{RO} + q_{RU} + (2q_B + q_G))\cos\theta \qquad 3.1$$

where
  f is the artificial friction coefficient,
  L is the conveyor length (pulley centre to pulley centre) (m),
  q_{RO} is the mass of the rotating parts of the upper idlers (kg/m),
  q_{RU} is the mass of the rotating parts of the lower idlers (kg/m),
  q_B is the mass of the belt (kg/m),
  q_G is the mass of the material transported (kg/m)
  g is the acceleration due to gravity (m/s^2) and
  θ is the angle of the conveyor to the horizontal.

If the angle of the conveyor is less than 18° then, in the method given in ISO 5048, the factor $\cos\theta$ is ignored.

Clearly a key factor in the success of this approach is selection of an appropriate value for the artificial friction coefficient, which we will consider later.

The secondary resistances are then expressed via a factor C, which, according to ISO 5048 is 'without risk of too serious an error'. This essentially corrects the length of the conveyor and is expressed in ISO 5048 as the sum of the main and secondary resistances divided by the main resistance. This factor C changes with the conveyor length and values are provided either in tabular form or graphically in the standards.

The slope resistance F_S (kN) can be calculated directly and accurately from basic mechanics and is given by

$$F_S = q_G.H.g \qquad 3.2$$

where
  q_G is the mass of the material transported (kg/m),

H is the vertical height through which the load is lifted from the point of loading to the point of discharge (m), and

g is the acceleration due to gravity (m/s^2).

For downhill conveyors this quantity will be negative and will have to be subtracted from the total resistance.

According to ISO 5048, the overall resistance F_U then becomes:

$$F_U = C.f.L.g(q_{RO} + q_{RU} + (2.q_{RB} + q_G)) + q_G.H.g + F_{S1} + F_{S2} \quad 3.3$$

where F_{S1} and F_{S2} are the special main and special secondary resistances respectively.

The standards give rules for the determination of the special main and special secondary resistances, which naturally are much lower than the main and special resistances. The chosen value for the artificial friction factor will depend on the condition of the installation and its maintenance. It may be as low as 0.017 for well-engineered and maintained installations or as high as 0.03 for installations operating in unfavourable conditions. For conveyors involving steep declines, lower values of the factor are applied (0.012 to 0.016). In addition, corrections can be applied to the artificial friction factor for temperature and speed of the conveyor. Temperature affects the viscosity of the lubricating grease in the idler bearings and the properties of the belt. Ketelaar and Davidson[3.10] report that the rolling resistance of idlers can double between +15 and −15 °C. However, they also comment that because of the large number of combinations of idler seal type, lubricating grease and belt construction materials, accurate determinations of temperature corrections are difficult.

The motor power required to move the conveyor is then given by:

$$P_M = \frac{F_U.\nu}{\eta} \quad 3.4$$

where

P_M is the motor power (W)

ν is the speed of the conveyor (m/s) and

η is the efficiency of the transmission.

3.4.2 'British Coal' approach

An alternative approach, which has been used by the former British Coal[3.5] and others,[3.11] considers the problem in a more general way and is, perhaps, easier to visualise. It assumes that the power needed is the sum of:

1. the power to move the empty belt
2. the power to move the load horizontally
3. the power to move the load vertically.

Design of belt conveyors 1 – power requirements and belt tensions

Of course, this approach and that outlined in Section 3.4.1 amount to the same thing, and indeed it will be seen that the simplifying assumptions in ISO 5048 do reduce them to very similar calculations as are given below. The power P_{eb} to move the empty belt is generally of the form:

$$P_{eb} = K.(L + L_0).W.s.g \text{ watts} \qquad 3.5$$

where
  K is a friction factor
  L is the belt length (m)
  L_0 is the length correction (m)
  W is the mass of the rotating parts of the idlers and of the belt (kg/m)
  s is the conveyor speed and is equivalent to v in BS 8438 (m/s)
  g is the acceleration due to gravity (m/s^2)

The power to move the load horizontally P_{lh} is given by:

$$P_{lh} = K.(L + L_0).F.g \text{ watts} \qquad 3.6$$

where
  F is the load conveyed (kg/s)

Thus we see that we have basically the same terms for moving the empty belt and the load horizontally in the formulae given in ISO 5048 and the above formulae 3.5 and 3.6, with:

$$K = f \quad \text{and} \quad (L + L_0) = C.L.$$

In reference 3.5, the method gives separate values for K to be used in the calculation of the power to move the empty belt and the power to move the load horizontally, on the grounds that tests on actual conveyor installations have shown that if this procedure is followed more accurate values for the power result. The values of K are given in that manual as 0.03 and 0.041, respectively for calculating the power to move the empty belt and the power to move the load horizontally. The factor C varies between 1.92 for a conveyor with a centre-to-centre distance of 80 metres to 1.03 for a 3000 metre long conveyor. The quantity L_0 is generally between 45 and 60 metres, which matches with values of C for conveyor lengths of around 2000 metres. As the calculation of the power to move the load vertically must be identical in the two approaches, this leaves the special resistances apparently unaccounted for in the 'British Coal' method. However, they are fully taken into consideration in the values of K and the corrected length. As with the friction factor, in reference 3.5 the length correction has two values, one for the empty belt and one for moving the belt horizontally. These values are $(1.45.L + 65)$ and $0.9.L$ respectively for the two cases.

The power to move the load vertically is given by

26 Belt conveying of minerals

$$P_{lv} = H.F.g \text{ watts} \quad\quad 3.7$$

where

H is the vertical height through which the load is raised (m)

The total power required is the sum of the three values determined, and the motor power is derived by dividing the total power by the efficiency of the motor and transmission.

3.4.3 CEMA approach

The CEMA approach is, similarly to that in ISO 5048, based on identifying the tension and power contributions from several friction mechanisms and components. The conveyor is broken into sections or 'flights' and the tension at any point such as the end of a flight is the sum of the tensions at the end of the previous flight and the tension changes along the current flight.

Basic method

The 'Basic' method applies to conveyors up to 800 feet long in a single flight, with a single loading point, fabric carcase belts, a single drive, belt speed up to 500 ft/min and belt tension up to 12 000 lbf. The tension T_e is given by:

$$T_e \leq W_m.H + 0.04(2.W_b + W_m)L \quad\quad 3.8$$

where

W_m is the distributed gravity load of the bulk material along length of belt (lb/ft)

H is the vertical height through which the load is lifted from the point of loading to the point of discharge (ft)

W_b is the distributed gravity load of the belt along length of belt (lb/ft)

L is the total conveyor path length tail to head (ft)

This equation appears to omit many of the factors found in the method given in ISO 5048. However, CEMA stress that this method is intended for use with basic conveyors only, that the method is conservative and that the tension to lift the bulk material forms a major part of the total tension.

Standard method

The 'Standard' CEMA method has historically been used for more complex conveyors than are appropriate for the Basic method: full details are given in the 5th edition of the CEMA book *Belt Conveyors for Bulk Materials*.[3.4] The Standard method applies to conveyors up to 3000 feet long in a single flight, with single or multiple loading points, fabric carcase belts, single or multiple drives, any belt speed and belt tension up to 16 000 lbf. The method is based, similarly to ISO 5048, on the identification and evaluation of all of the various

Design of belt conveyors 1 – power requirements and belt tensions

forces acting on the conveyor belt, such as:

- the force to lift or lower the material being transported
- the force required to overcome the frictional resistance of the conveyor components, drive and accessories
- the force required to overcome the frictional resistance of the material being conveyed
- the force required to accelerate the material being conveyed as it is fed onto the belt.

The basic equation for calculating the effective tension is:

$$T_e = L.K_t(K_x + K_y.W_b + 0.015W_b) + W_m(L.K_y + H) + T_p + T_{am} + T_{ac} \qquad 3.9$$

where

W_m, H, W_b and L are as previously defined for the Basic method,
K_x is the idler resistance factor (lbf)
K_y is the belt resistance factor (dimensionless)
K_t is the temperature resistance factor (dimensionless)
T_p is the tension due to belt flexure round pulleys and the pulley bearing resistance (lbf)
T_{am} is the tension due to the force to accelerate the material (lbf)
T_{ac} is the tension from accessories (lbf)

The factor K_x is dependent upon the belt mass, the material mass, the idler spacing and the idler type. The factor K_y is also dependent upon the belt and material masses and the idler spacing, and also on the slope angle of the conveyor. We can, of course, identify the same elements in this equation as the one in ISO 5048, since the elements of the conveyor are the same. However, in the Standard CEMA method separate factors are used for the frictional resistances of the belt and idlers, the idler mass is not used and there is no length correction coefficient. In ISO 5048 the single artificial friction factor is applied to the masses of idlers, belt and load. As in ISO 5048, the method has scope for including various corrections and additional minor factors where necessary.

Universal method

The CEMA Universal method, introduced in the sixth edition of *Belt Conveyors for Materials Handling*, is very detailed, but operates on the same principles as the Standard method. It is applicable to all conveyors. The equation for the effective tension in the belt has three terms:

1. an energy term
2. a term relating to the main resistances
3. a term relating to point resistances.

The first of these accounts for:

- the change in belt tension to lift or lower the material and belt
- the tension added in accelerating material to belt speed.

The second includes changes in tension due to:

- the belt sliding on chute skirtboard seals
- idler seal friction
- idler load friction
- visco-elastic deformation of belt
- idler misalignment
- drag due to slider beds
- bulk materials sliding on skirtboards
- bulk materials moving between the idlers.

And the third includes changes in tension due to:

- belt bending on the pulley
- pulley bearings
- belt cleaners
- discharge plough.

All apply to any single flight or pulley.

The Universal method gives details of how each of the individual contributions is calculated. Further detail is beyond the scope of this book and the reader should refer to the sixth edition of the CEMA publication *Belt Conveyors for Materials Handling*.

To aid calculation, some standards provide tables of values of, for example, cross-sectional area of material conveyed, mass per unit length of rotating parts and conveyor belts, power to move the load horizontally and vertically. Other standards may provide similar information graphically.

3.4.4 Comparison of power calculations

Kirk[3.12] in his paper on the effect of drive configurations on belt performance gives a comparison of the powers for two installations calculated by the CEMA and ISO 5048 methods. As the paper was written in 1998 it is assumed that the CEMA method is the Standard method. From the information in his paper and from a knowledge of the installations concerned we have calculated the powers using the British Coal *Belt Conveyor Handbook*[3.5] and the formulae given in the *Fenner Dunlop Technical Manual*[3.11] to provide a more complete comparison. For the CEMA and ISO methods we do not have details of the assumptions made, such as the value of the artificial friction factor, but for the Fenner method we give power calculations for friction factors of 0.03 and 0.02. For the British Coal method we have used the tables provided in the handbook, which have

Design of belt conveyors 1 – power requirements and belt tensions

Table 3.2 Power calculation comparisons

	Installation A		Installation B	
Horizontal length (m)	1420		2530	
Vertical lift (m)	200		110	
Belt speed (m/s)	2.6		4	
Belt width (mm)	1200		1350	
Installed power (kW)	895		1490	
Tonnes/h	0	1000	0	2400
Power CEMA (kW)	75	715	204	1352
Power ISO 5048 (kW)	91	723	225	1395
Power Fenner (0.03) (kW)	113	777	337	1559
Power Fenner (0.02) (kW)	75	699	224	1280
Power British Coal (kW)	109	797	407	1716

been calculated using standard values of parameters such as idler types for given belt widths, idler and belt masses, and the friction factors and corrected lengths given in Section 3.4.2 above. Because the installations concerned were not using one of the British Coal standardised belt types, the belt masses used are likely to underestimate the actual values. The results of the calculations are shown in Table 3.2.

It is interesting to note the similarity in the values of power calculated by the CEMA and ISO methods and the similarity of those values to the values derived from the Fenner method when a friction factor of 0.02 is used. The values derived by the British Coal method appear very conservative, but this is only to be expected given the nature of the installations, which were in coal mines, and the severity of the expected operating conditions. For Installation B the power calculated from the British Coal tables is greater than the actual installed power, but this may be due to the assumptions made in the standard tables in that manual.

3.5 Belt tensions

3.5.1 Basic consideration of tensions

Having determined the power required to move the conveyor and the belt speed, it is then possible to calculate the tensions in the belt. The magnitude of the tension is needed so that the correct type of belt can be selected for the duty.

The power P transmitted to the belt by the drive pulley is given by

$$P = T_e . s \qquad 3.10$$

where
 T_e is the effective tension at the pulley, and
 s is the belt speed.

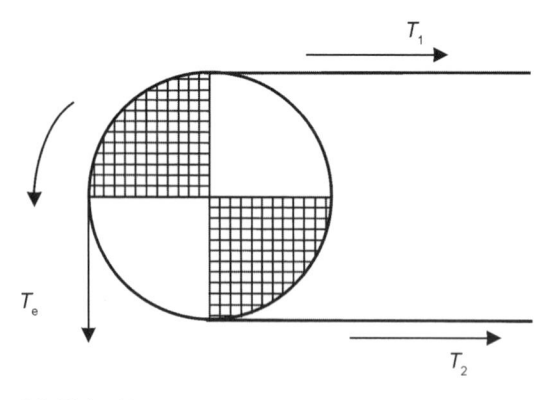

3.3 Plain drive.

By definition $T_e = T_1 - T_2$, where T_1 is the carry side tension and T_2 is the slack side tension (Fig. 3.3). Using the additional relationship derived by Swift[3.13]

$$\frac{T_1}{T_2} = e^{\mu\theta} \qquad 3.11$$

where

e is the base of Naperian logarithms,
μ is the coefficient of friction between the belt and the pulley, and
θ is the angle of wrap of the belt around the pulley (radians),

values for T_1 and T_2 can be calculated. This last relationship represents the limiting tension ratio when the belt is about to slip. The coefficient of friction, the angle of wrap and the drive arrangement can all affect the belt tensions. In some design standards and guides the quantity $1/(e^{\mu\theta} - 1)$ is defined and referred to as the 'drive factor'. It is derived and used as follows:

Since $\dfrac{T_1}{T_2} = e^{\mu\theta}$,

$$T_1 - T_2 = T_2(e^{\mu\theta} - 1),$$

Then $\dfrac{T_2}{T_1 - T_2} = \dfrac{1}{e^{\mu\theta} - 1} = \dfrac{T_2}{T_e},$

and $\quad T_2 = T_e \left(\dfrac{1}{e^{\mu\theta} - 1} \right) \qquad 3.12$

Let $\dfrac{1}{e^{\mu\theta} - 1} = K$

Then since

$$T_e = T_1 - T_2$$

Design of belt conveyors 1 – power requirements and belt tensions

$$T_e = T_1 - T_e\left(\frac{1}{e^{\mu\theta} - 1}\right) = T_1 - T_e.K$$

and $\quad T_1 = T_e(K+1)$ 　　　　　　　　　　　　　　　　　　3.13

Since the power required is fixed, and this fixes the value of the effective tension T_e, if the speed is also fixed, decreasing the value of the drive factor decreases the value of T_1. This may allow a lighter, less expensive belt to be used and reduce the amount of wear in the system. The drive factor is dependent upon the coefficient of friction μ between the belt and the drive pulley, and on the angle of wrap θ of the belt around the pulley. Increasing either or both of these quantities will result in a lower value of T_1. To illustrate this, assume that the coefficient of friction is 0.25 and that the angle of wrap is 180°, i.e., a plain drive. Then $e^{\mu\theta} = 2.1933$ and the drive factor becomes 0.838 and $T_1 = 1.838.T_e$. If now we use a snub pulley to increase the angle of wrap to 230° then $e^{\mu\theta}$ becomes 2.718 and the drive factor becomes 0.5821. Then $T_1 = 1.5821.T_e$. It is worth noting at this stage that if the coefficient of friction is higher in practice than the design figure, the real value of T_1 will be lower than the calculated value.

3.5.2 Complex drives

Changing the design more radically can reduce the value of T_1 further. Figure 3.4 illustrates a dual-pulley geared tandem drive. In this arrangement the pulleys are coupled together by gears and the power is shared unequally between them through the gearing. There may be one or more motors mounted on the main drive pulley.

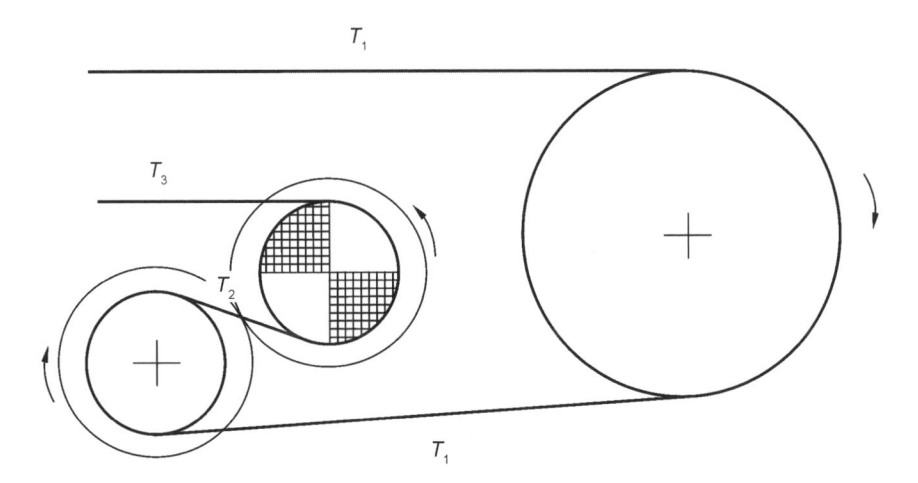

3.4 Dual-drum geared tandem drive.

Assuming the same angle of wrap on the two pulleys,

$$\frac{T_1}{T_2} = \frac{T_2}{T_3} = e^{\mu\theta} \qquad 3.14$$

$$T_3 = \frac{T_2^2}{T_1}$$

$$T_2 = \frac{T_1}{e^{\mu\theta}}$$

Substituting

$$T_3 = \frac{T_1}{e^{\mu\theta}.e^{\mu\theta}}$$

$$T_1 - T_3 = \frac{P}{S} = T_e,$$

and $$T_1 = \frac{P}{S}\left(\frac{(e^{\mu\theta})^2}{(e^{\mu\theta})^2 - 1}\right) \qquad 3.15$$

Substituting the values of coefficient of friction of 0.25 and angle of wrap of 230° gives $T_1 = 1.1565.T_e$. In another arrangement used, each drive pulley is powered by a separate motor, i.e. not geared, the dual-pulley dual-motor drive (Fig. 3.5). With natural load sharing the distribution of tensions and loads between the two pulleys would be as for the dual-pulley geared tandem drive above and it can be shown that one motor would take approximately 70% of the total load. As Gilbert[3.14] pointed out, this requirement for motors of different sizes would be inconvenient in practice and in fact was not followed in the coal mining industry. He noted that dual-pulley dual-motor drives had been run

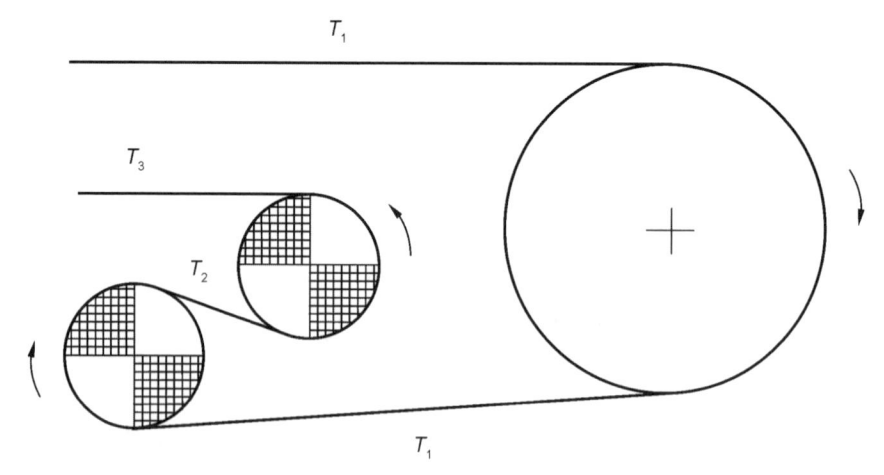

3.5 Dual-drum dual-motor ungeared drive.

Design of belt conveyors 1 – power requirements and belt tensions

successfully with equal motors, giving savings in terms of rationalisation, standardisation and spares. He calculated that with natural load sharing, and assuming $T_e = 100$ arbitrary units

$$T_1 = 116, \ T_2 = 42.5 \text{ and } T_3 = 16,$$

Hence $T_1 - T_2 = 73.5$ and $T_2 - T_3 = 26.5$.

However, with equal load sharing

$$T_1 - T_2 = T_2 - T_3 = \frac{T_e}{2} = 50$$

$$\frac{T_2}{T_3} = e^{\mu\theta} = 2.72$$

$$T_2 = 2.72 \cdot T_3,$$

Thus $T_2 - T_3 = 2.72 \cdot T_3 - T_3 = 50$
$T_3 = 29$

Then $T_2 - T_3 = 50$
$T_2 = 79$

and $T_1 = 129$

Increasing T_3 in the ratio 16 to 29 the load is shared equally between the two motors, but T_1 is increased in the ratio of 116 to 129. Gilbert states that in practice the power demand for each motor can be balanced by increasing T_3 and counterbalancing the difference in the motor characteristics by 'trimming' the fluid couplings. (These couplings are interposed between the prime mover and the pulley to avoid imparting high dynamic stresses to the belt under conveyor start conditions and are discussed in the following chapter.) This example illustrates the possibility of changing tensions to optimise the conveyor design.

Other more complex drive systems are in use; for example, the four motor, four drive pulley ungeared (Fig. 3.6), the dual-pulley ungeared primary plus

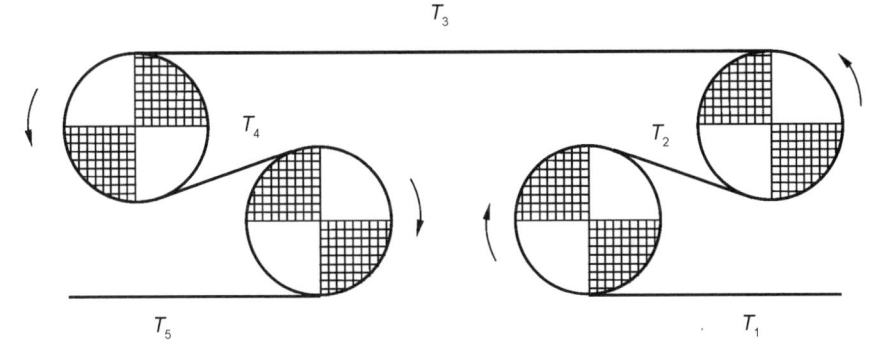

3.6 Four motor, four drive pulley ungeared drive.

34   Belt conveying of minerals

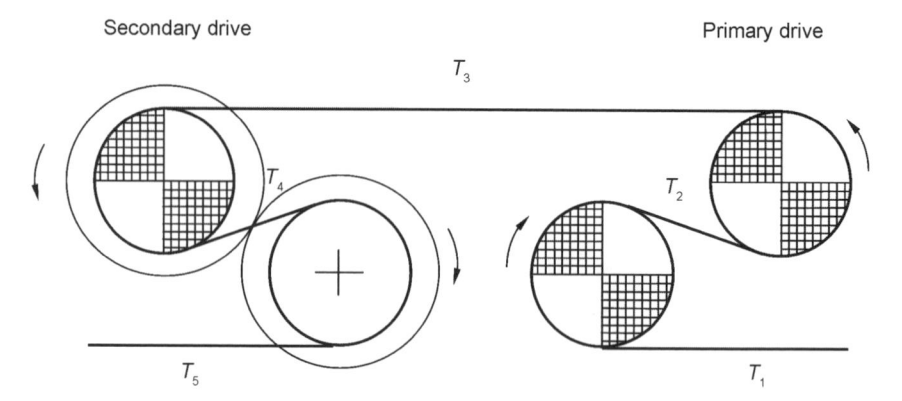

3.7 Dual-pulley ungeared primary plus geared tandem secondary drive.

geared tandem secondary (Fig. 3.7) and the two geared tandems (Fig. 3.8). For these forms of drive the same principles for calculating the tensions as are illustrated above apply. It should be noted, however, that the limiting slip condition can occur at any of the drive drums and it is therefore necessary to check each case. The consequences of belt slip can be severe and even include the generation of fires (see Chapter 8). It is therefore usual to err on the side of caution and to assume conservatively low values of friction coefficient, to use lagging on the drive drums to increase the friction coefficient or to increase the angle of wrap to above the minimum required.

On complex drive systems account may need to be taken of the elastic and visco-elastic properties of the belt. Two effects arise. Because the belt is elastic it changes its length as the tension changes. Thus an element of belt going onto a drive pulley at high tension T_1 is longer than when it leaves the drum at lower tension T_2. Since the drum surface is travelling at constant speed the belt must move relative to the drum surface. This phenomenon, known as creep, is

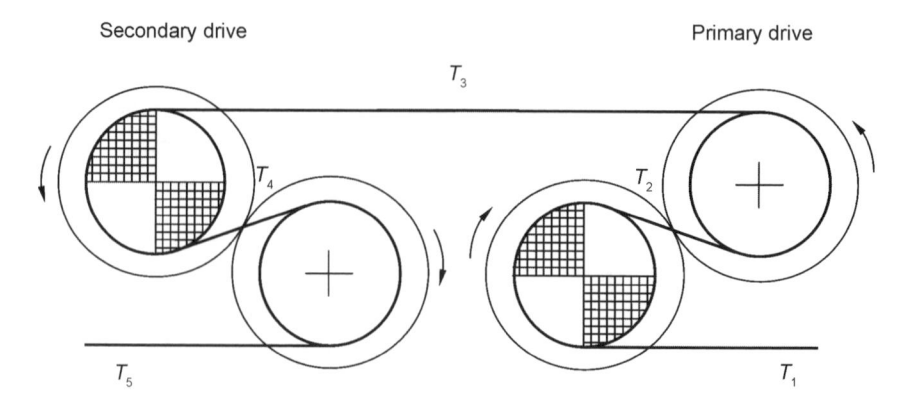

3.8 Two geared tandems drive.

explained fully by Firbank.[3.15] Further to this, in addition to the tensile stress imparted to the belt by the driving pulley there will be a bending stress caused by the belt passing round the pulley. This bending stress will not be uniform across the belt and the combination of these stresses results in a complex stress distribution through the belt thickness. Barclay[3.16] used strain gauges fitted into plied belts to quantify these stress distributions, revealing a zone of sharply increasing stress due to bending as the belt went onto the pulley, a constant stress zone and a zone of gradually decreasing stress as the belt travelled round to leave the pulley. He also showed that the neutral axis was close to the centre of the belt. Barclay's tests were made at very low speeds where visco-elastic effects would not be important, but in practice the recovery of strains imposed on the belt as it passes round the drive is important. Gilbert[3.14] remarks on the existence of a 'pit rule' that required the length of belt between the primary and secondary drives to be greater than '2 ft/ft/s of belt speed' (2 m per m/s of belt speed). Hence for a belt travelling at 2.5 m/s the drums would need to be separated by five metres. Kirk[3.12] reports that traditionally the British Mechanical Handling Equipment Association (BMHEA) recommends that the distance between drive pulleys should be 3.5 m or the distance travelled by the belt in one second, whichever is the greater. However, he remarks that these criteria do not allow for the effect of the belt travelling round snub pulleys. According to Gilbert, the distance between the driving pulleys prevents the effect of creep 'scuffing' between the belt and the driving pulleys, while the BMHEA require this distance to allow the belting to recover from the elongation resulting from the difference in tension between the belt coming off the primary pulley and going onto the secondary pulley. The following case history demonstrates the importance of taking these elastic and visco-elastic effects into account on complex drives. It is taken from events with which the authors were involved and which are reported on fully by Kirk.[3.12]

3.5.3 Case history

Because of their construction, steel cord carcase belts can be made with very high strengths and have very low stretch under load. This means that they can be used on high power installations used for conveying high tonnages. However, they can be prone to ripping and to cover damage that leads to corrosion of the cords that, in turn, leads to loss of strength, Following experience of such problems, British Coal sought to source a high strength, low stretch, textile carcase belt as an alternative to steel cord belts. The alternative fabric carcase belt would need to match the high strength and low stretch characteristics of the steel cord belts as far as possible. A new 22.5 mm thick, 3150 N/mm (18 000 lb/in), solid woven, PVC impregnated belt with nitrile rubber covers over PVC covers was tried on the two installations detailed in Table 3.2 above. (For details of belt types and constructions see Chapter 5.) Both installations were double pulley ungeared

36   Belt conveying of minerals

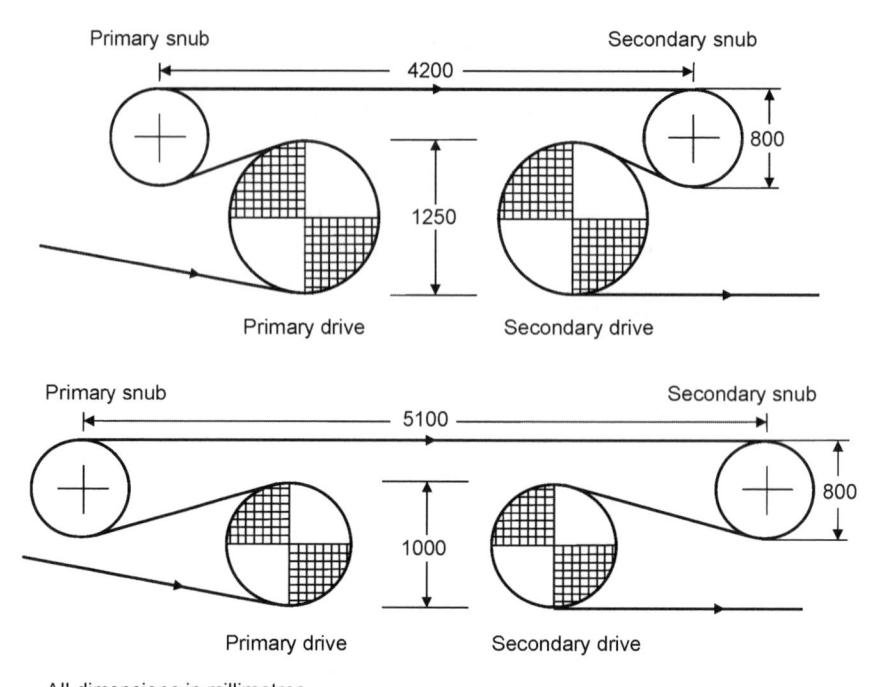

3.9 Drive configurations for Installations A (upper) and B (lower).

drives of the 'omega' configuration shown in Fig. 3.9, although the sizes of the pulleys and the distances between them differed.

The new belt type was put to work initially on Installation B, and apart from some early problems with minor blistering, which was put down to contamination during manufacture, the belt worked satisfactorily throughout the life of the installation. However, within two or three weeks of being put to work, the belt on Installation A began to suffer blistering of the rubber cover on the pulley side. The blisters were initially 15–50 mm in diameter and standing up about 25 mm from the belt surface. Although the blistered areas were repaired on site, continued running of the conveyor resulted in further deterioration, such that after eleven months running 80 to 90% of the surface of the belt on the pulley side was damaged. The bonding failure was found to be at the interface between the PVC cover and the PVC-impregnated carcase. The problem was initially thought to be due to manufacturing defects, but despite all attempts by the belt manufacturer to improve the bonding, no improvement in belt performance was seen, and further running produced whole areas of delamination of the cover.

The belting had originally been tested on a four-pulley laboratory rig (see Section 7.5 and Fig. 7.5 of Chapter 7) with 1000 mm diameter pulleys. Tests were repeated with 800 mm diameter pulleys and produced the same type of damage as on Installation A. The belt design was modified to give greater

Design of belt conveyors 1 – power requirements and belt tensions   37

Table 3.3 Drive configurations in Installations A and B

Design aspect	Installation A	Installation B
Primary drive pulley diameter (mm)	1250	1000
Angle of wrap (degrees)	229	215
Primary snub pulley diameter (mm)	800	800
Angle of wrap (degrees)	208	215
Distance between snub pulleys (mm)	4200	5100
Secondary snub pulley diameter (mm)	800	800
Angle of wrap (degrees)	230	205
Secondary drive pulley diameter (mm)	1250	1000
Angle of wrap (degrees)	229	210

elasticity, and was successfully run on the test rig with 800 mm pulleys. However, when put on Installation A the new belt also failed by delamination. Detailed investigations were then made into the dynamics of the drives on the two installations to determine what the differences were and whether these could explain the failure of the belt on Installation A. Details of the two drive configurations are shown in Table 3.3.

Values of power requirement, tensions and tendency to slip calculated for both installations showed no areas of concern. Analysis of the distances between the various pulleys in the drives initially showed that the drive on Installation A was much 'tighter' than that on Installation B. However, taking into account the differences in belt speeds there was little to choose between the drives in terms of 'flight' times between the different pulleys. There was one exception to this and that was the distance between the secondary snub and the secondary drive pulleys, which was much shorter for Installation A than for Installation B and gave a much shorter flight time, 0.12 seconds compared to 0.29 seconds. On-site measurements of the performance of the conveyor, the speeds of the belt through the drive at various points and the respective tensions again gave little concern, but these measurements, together with laboratory tests to determine the rate of recovery of the belt from shear, allowed the degree of recovery of the belt from strain induced by shear to be determined as the belt passed round the drives. It was clear from these measurements that there was considerable locked-in strain in the belt as it passed round the drives. Changes in cover elongations as the belt passed through the drives were very similar for Installations A and B and it was concluded that Installation B was close to the limit of operation for this type of belt. The high locked-in strains in the belt, arising from its visco-elastic properties and the short distance between the secondary snub and the secondary drive on Installation A, were probably the cause of the belt failure. A solution to the problem would have been to increase the distance between the two pulleys concerned, but this was not implemented since it would have taken the secondary snub outside the drive module.

This case history demonstrates the importance of taking account of belt properties when designing drives, particularly when new belt or drive designs are employed, and not relying solely on traditional rules of thumb or 'pitmatics'.

3.5.4 Minimum tension

The minimum tension in the system must not only prevent belt slip, but also has to be such as to limit to an acceptable level the belt sag, which is defined as the vertical displacement between idler sets. A value of 2% of the idler pitch (spacing) is usually taken as appropriate. The sag tension T_s in newtons is calculated according to formulae of the following form (refs 3.2, 3.3, 3.5):

$$T_s = \frac{a_r \cdot g \cdot q_B}{8(h/a_r)} \qquad 3.15$$

for the return side and

$$T_s = \frac{a_c \cdot g \cdot (q_B + q_G)}{8(h/a_c)} \qquad 3.16$$

for the carrying side, where
  a_r is the return idler pitch (m)
  a_c is the carrying idler pitch (m)
  q_B is the mass of the belt (kg/m)
  q_G is the mass of the material conveyed (kg/m)
  h is the allowable sag (m)
  g is the acceleration due to gravity (m/s²)

We shall discuss the choice of idler spacing in Chapter 4.

Some standards advise that the tensions in the system should be displayed graphically. This allows the distribution of tensions in the system to be examined and is particularly useful for complex conveyor profiles where several changes in gradient may be present and the profile may include downhill sections. It enables the maximum tension in the system, the magnitude of the external tensioning device on the slack side and the effect of acceleration and braking to be determined. By examining all of the load cases for complex conveyors the designer may reveal whether circumstances exist in which the conveyor becomes regenerative, i.e., the conveyor is supplying power to the drive as it might be on a downhill section, and therefore needs to be braked.

At this stage we have dealt with the basic principles behind the determination of conveyor belt width and speed, belt capacity, power requirements and belt tensions. For full treatments of these and other relevant factors the reader is advised to consult references 3.1 to 3.4. The design process will, of course, be iterative, but having got this far it would be possible to make an initial choice of belt strength. Some design methods[3.5] advise on further refinements to the calculations of tensions, such as determining whether a correction to T_1 and T_2

Design of belt conveyors 1 – power requirements and belt tensions

needs to be made for the effect of the mass of the belt if the conveyor is on an incline. Similarly, account may need to be taken of the effects of acceleration and braking on tensions. These matters are outside the scope of this book, which is intended to provide the basics of the design process.

Manufacturers of conveyor belts recommend maximum values of running tension for their belts, expressed in newtons per millimetre width or pounds per inch width in the USA. These values are much lower than the tensile strength of the belts and incorporate a 'service factor' or 'factor of safety' to allow for various operational considerations. We shall discuss this further in Chapter 5.

We have stressed previously that much of conveyor design is concerned with the control and management of tension in the conveyor belt. However, we have not considered circumstances where the tension at different points across a belt may differ and may change. This is dealt with in the following section.

3.6 Tension changes over the belt width

The perceptive reader may be concerned at this point that so far nothing has been said about the fact that the belt must pass from a troughed shape to a flat one and vice versa on its journey round the conveyor and whether this matters in terms of tension in the belt. Equally, we have not considered changes in gradient or direction, which must cause vertical or horizontal curves to be induced in the conveyor. The reader would be right in thinking that all of these are important as they cause differential tensions to arise in the belt and must be carefully controlled. The change from troughed to flat and vice versa is referred to as the *transition* and the distance over which it takes place is the *transition distance*.

3.6.1 Transition distance

Consideration of the basic geometry of the transition under steady-state running reveals that there must be a difference in the distance travelled by the centre and the edges of the belt. This causes a difference in the tensions in these two locations. The transition distance needs to be such that the edge tension does not rise so high that the elastic limit of the belt is exceeded and damage is caused to the belt edges, or fall below zero at the centre and so cause buckling of the belt. For non steady-state conditions, such as starting and stopping, the edge stresses will be higher still.

For a given conveyor trough cross-section and position of the terminal pulley relative to the centre idler, the minimum transition distance can be calculated once the recommended maximum running tension and the elastic modulus of the belt are known. ISO 5293:2004[3.17] provides formulae for the calculation of the belt edge tension and the minimum transition distance. The formula for the transition distance is of the form:

$$L = \frac{h}{\sin \lambda} \left[\frac{M}{\Delta T} (1 - \cos \lambda) \right]^{0.5} \qquad 3.17$$

where

L is the transition distance (m)
h is the distance the belt edge rises or falls in the transition (m)
λ is the troughing angle
M is the elastic modulus measured at the recommended maximum belt-to-belt joint tension (N/mm)
ΔT is the induced belt edge stress (N/mm).

This standard shows the stress distribution across a belt at the transition and gives methods of calculation for the maximum edge tensions and the tension at the centre of the belt for the avoidance of buckling at the belt centre. The calculation of transition distance needs to be made for each case and the longer of the two results taken as the transition distance. The standard indicates that the maximum allowable edge stress should be agreed with the belt manufacturer, but gives guidance that for textile belts it may be 180% of the maximum running tension and 200% for steel cord belts.

The transition distance may be reduced if the pulley is raised relative to the centre idler because this reduces the edge stress, but should be lengthened if the pulley is lowered because this raises the stress in the belt edges.

DIN 22101[3.3] also provides methods of calculation for transition distances based on similar geometric principles. However, it provides different formulae for textile and steel cord belts. Whereas ISO 5293:2004 makes no reference to any need for special measures with steel cord belts, DIN 22101 remarks that because these belts stretch so much less than textile belts particular care is needed over the accuracy of the calculation of transition distances. It is insistent that using the same calculation methods for steel cord belts as are used for textile belts can lead to important inaccuracies. It provides guidance on the determination of transition distance if no other relevant information is available, using the relationship:

$$L = c.h \qquad 3.18$$

where

L is the transition distance
h is the distance the belt edge rises or falls in the transition
c is a constant equal to 8.5 for nylon-cotton textile carcase belts and 14 for steel cord belts

The calculation of the transition distance for steel cord belts takes a similar form to that for textile belts, but introduces empirical factors that allow for the fact that in steel cord belts the differences in length and strain produced by the transition are equalised over much longer distances than in textile belts.

Design of belt conveyors 1 – power requirements and belt tensions   41

The 5th edition of the CEMA book *Belt Conveyors for Bulk Materials*[3.4] provides guidance for the determination of transition distances in tabular form. The tables show the transition distance as a multiple of belt width for textile and steel cord belt types respectively, as a function of troughing angle and percentage of rated belt tension. One table provides these data for the case when the top of the terminal pulley is positioned level with the top of the centre idler (full trough depth) while the other table provides this information when the top of the terminal pulley is at half the trough depth above the top of the centre idler (half trough depth). This is a much less complex method of determining the transition distance than is given in either the ISO or DIN standards. Values for the elastic modulus for the belts must have been assumed: no explanation is given as to how the figures presented were derived.

3.6.2 Vertical and horizontal curves

Vertical curves may be convex or concave. A convex curve is defined as one in which the centre of curvature of the curve lies below the conveyor. With either convex or concave curves there are certain important design rules that need to be followed to avoid damage to the belt or other equipment. Because the belt is troughed, the centre and edges will be required to conform to different radii on a curve and it is therefore important that the radius of curvature is sufficiently large to avoid belt stresses being either so high as to cause damage, or so low as to cause buckling of the belt. With convex curves the edges of the belt can become overstressed while at the centre of the belt the stress can fall below zero. In addition the downward component of the belt tension can result in overloading of the idlers and reduced spacing is sometimes employed to avoid this. In the case of concave curves it is the belt centre that may become overstressed and the edges where the stress reduction may result in buckling. The belt may also tend to lift off the idlers and shed the load.

Some standards and manuals provide equations for calculating the minimum radius of curvature. These tend to be empirical and derived from experience. For example, reference 3.5 gives, for concave curves on start up:

$$R = \frac{1051.T.A}{M.g} \qquad 3.19$$

where
 R is the radius of curvature (m)
 T is the belt tension at the change in gradient (kN)
 A is the acceleration factor
 M is the belt mass (kg/m) and
 g is the acceleration due to gravity (m/s^2)

The value of the factor A varies depending upon the type of coupling in the drive system, varying from 1.3 for acceleration limiting control couplings to 1.6 for traction couplings. Where no coupling is present, the factor A is equal to the ratio (available motor torque: full-load torque). This reference also gives formulae for calculating belt edge and centre stresses and potential belt lift for concave curves, while for convex curves it gives formulae for belt edge and centre stresses and idler loadings. DIN 22101 gives a warning that accurate values of strains for vertical curves require precise and expensive calculation methods. However, it provides an approximate calculation method that follows from the formulae given for transition distances. It uses geometries to determine strains and the elastic modulus values needed for the transition distance calculation to determine tensions. Limiting values for the strains at the centre and edges are provided, based on the vertical distances of these locations from the neutral axis of the belt cross section. CEMA[3.4] similarly provides comprehensive information for the design of both convex and concave curves.

Normally, horizontal curves would be avoided and indeed DIN 22101 warns that only very small horizontal deviations are possible. However, in recent times some conveyor companies have found it necessary to incorporate horizontal curves in their designs. This has been made possible in the main by improved control of belt tensions through the use of the so-called 'soft start' technology and the use of intermediate drives known as booster drives in the system and improved belt tension sensing equipment, although improvements in belt technology have played their part. These aspects are discussed further in Chapter 4.

3.7 Concluding remarks

In this chapter we have looked at the basic factors involved in the calculation of the capacities, power requirements and belt tensions for belt conveyors. We have seen that although the design calculations can appear relatively straightforward, there are pitfalls that can trap the unwary. We have tried to give references to suitable guides to conveyor design and show the differences in approach that exist. We have also tried to demonstrate that operational experience gained over many years is invaluable and needs to be preserved, but that radical changes, e.g., in belt properties, can take us outside the fields of application of our rules of thumb. In the next chapter we shall consider the hardware that comprises a belt conveyor.

3.8 References

3.1   ISO 5048:1989 'Continuous mechanical handling equipment – Belt conveyors with carrying idlers – Calculation of operating power and tensile forces'.
3.2   BS 8438:2004 'Troughed belt conveyors – Specification'.
3.3   DIN 22101 'Continuous conveyors – Belt conveyors for loose bulk materials – Basics for calculation and dimensioning'.

Design of belt conveyors 1 – power requirements and belt tensions

3.4 Conveyor Equipment Manufacturers' Association (CEMA) *Belt Conveyors for Materials Handling*, 5th Edition, 1997.
3.5 *Belt Conveyor Handbook*, Notes for Guidance NG/6 British Coal.
3.6 Fayed M E and Skocir T S *Mechanical Conveyors – Selection and Operation* Technomic Publishing Company Inc. Pennsylvania USA 1997.
3.7 CEMA 550:2003 'Classification and Definition of Bulk Materials'.
3.8 ISO 251:2003 Conveyor belts with textile carcass – Widths and lengths.
3.9 Streets H *Sutcliffe's Manual of Belt Conveying* W & R Chambers Ltd 1956.
3.10 Ketelaar J P I and Davidson P J 'Improving the efficiency of conveyors used for the transport of minerals in underground and surface mines' *Mining Technology* Vol. 77 No. 883 March 1995.
3.11 *Fenner-Dunlop Technical Manual* Fenner-Dunlop Conveyor Belting Europe Limited, Hull UK.
3.12 Kirk A 'Drive configuration effects on conveyor belts' *Proceedings of Beltcon 8 Conveying of materials and problems encountered* October 1998 Johannesburg.
3.13 Swift H W 'Power transmission by belts: an investigation of fundamentals' *Proc Inst Mech Eng* Nov 1928.
3.14 Gilbert S Discussion to 'A laboratory study of the dual-drum multimotor conveyor drive' by Firbank T C *The Mining Engineer* No 140 May 1972.
3.15 Firbank T C 'A laboratory study of the dual-drum multimotor conveyor drive' *The Mining Engineer* No 140 May 1972.
3.16 J T Barclay *Conveyor Belting Research. A monograph on the work carried out on conveyor belting at the Mining Research Establishment of the National Coal Board during the period 1950–1966*, UK National Coal Board 78pp.
3.17 ISO 5293:2004 'Conveyor belts – Determination of minimum transition distance on three idler rollers'.

4

Design of belt conveyors 2 – hardware (idlers, structure, pulleys, drives, tensioning devices, transfer points and belt cleaning)

4.1 Introduction

The perceptive reader will have noticed that in the previous chapter we have discussed idlers and pulleys and in particular have used quantities such as the mass of the moving parts of idlers and the spacing of top and bottom idlers in the calculations of powers and tensions. Yet thus far we have not discussed the actual design and construction of these items or the way in which they fit into the conveyor system. This has been done deliberately since the broad considerations of belt capacity, power requirements and belt tensions do not require detailed knowledge of idler, structure or pulley design. We shall now consider these aspects, together with the design of drives, tensioning mechanisms, materials transfer points and belt cleaning devices.

4.2 Conveyor idlers

4.2.1 Idler sets

In the introduction to Chapter 3 we defined some of the common terms associated with conveyors. In the definition of carrying idlers we said that there were usually three idlers to a set, one centre idler and two wing idlers. We should point out that five-idler sets exist, with a horizontal centre idler and two wing idlers at either side, the outer being inclined at a steeper angle to the horizontal than the inner. However, the economics of five-idler sets seem questionable and they are not common. With three-idler sets the horizontal axis of the centre idler may be in line with the axes of the two wing idlers (in-line idler sets) (Fig. 4.1) or it may be offset (staggered idler sets) (Figs 4.2 and 4.3). This offset or staggered centre idler design allows the ends of the centre idler to overlap those of the wing idlers and this removes the gap between the centre and wing idlers that must exist if the axes of three idlers are all in line. This has the advantage of removing a potential pinch point that can nip the belt and cause damage to it. These staggered idler sets have been in common use in the UK coal mining industry and are said to give better control of the belt by the centre idler.[4.1] In the Introduction to Chapter 3 we

Design of belt conveyors 2 – hardware   45

4.1 In-line idler set (by courtesy of ATH Resources Ltd).

also defined return idlers, stating that there is usually a single flat return idler. However, there may be two return idlers in a 'V' shape, each set at an angle of up to 10° to the horizontal. To aid belt tracking the carrying wing idlers and the vee return idlers may be tilted forward by up to 3° for carrying idlers and up to 1° for vee return idlers. BS 8438$^{4.2}$ gives details of geometrical arrangements of idlers and the clearances needed between idlers in a set and between idlers and supporting structure to minimise damage to the belt.

Another type of idler that is said to aid belt control and alignment is the garland or catenary idler, which may come in two, three or five roller sets.

4.2 Staggered idler set (by courtesy of ATH Resources Ltd).

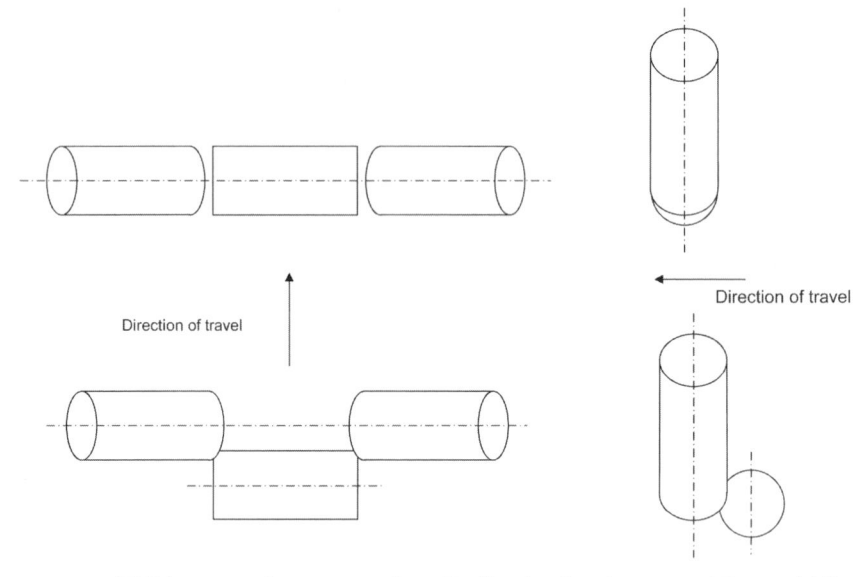

46   Belt conveying of minerals

4.3 Diagrammatic representation of in-line (top) and staggered (bottom) idler sets.

Garland idlers are suspended from the conveyor support framework and have flexible connections between the individual idlers so that the shape of the trough formed can follow, say, an off-centre load caused by large lumps (Fig. 4.4). As with the rigid idler sets, the two-roller idlers are used on the return strand. There are various ways in which carrying and return idlers of all types may be supported: the choice depends on the duty to which the conveyor is put, but is

4.4 Garland idler set (by courtesy of ATH Resources Ltd).

Design of belt conveyors 2 – hardware   47

4.5 Conveyor structure suspended from mine tunnel roof (former NCB photograph – published by kind permission of Department for Business, Enterprise and Regulatory Reform).

often in the form of a steel framework, which in coal mines may be suspended from the roof (Fig. 4.5).

In certain parts of the conveyor, such as loading points, the normal idlers may be substituted by impact idlers, which are designed to allow some resilience in the support of the belt to avoid belt damage. Impact idlers are most often rubber-covered steel but other designs, such as those involving a series of resilient discs on a shaft, are in use. One design is shown in Fig. 4.6.

4.2.2 Design of idlers

A conveyor idler roller basically consists of a tube that turns on a shaft via rolling element bearings fitted into end plates attached to the tube. Seals are

4.6 Impact idlers (by courtesy of ATH Resources Ltd).

Belt conveying of minerals

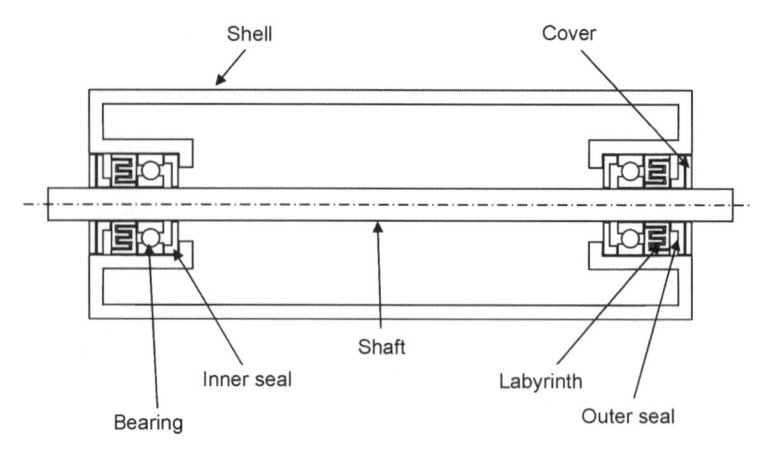

4.7 Diagrammatic representation of a section through a conveyor idler roller.

fitted to prevent the ingress of contaminants to the bearings (Fig. 4.7). The detailed design and assembly of these components affects both the resistance of the idler to rotation – and hence the power requirement for the conveyor – and the life of the idler. Ketelaar and Davidson[4.3] have identified the following factors that affect the resistance of the idler to rotation:

- roller and bearing rating and size
- type of sealing
- type and quantity of lubricant
- manufacturing accuracy and consistency
- rotational speed
- ambient temperature.

They also remark that resistance increases with rotational speed, and that the better the sealing arrangement, the higher the resistance to rotation, with labyrinth seals being the least resistant and face seals the most effective. Some organisations may specify figures for rotational torque of idlers. British Coal in its Specification for heavy duty idlers stated that static torque should not exceed 1.2 Nm, the dynamic torque after one hour's running should not exceed 2.0 Nm at 680 rev/min and after one hour's standing, the dynamic torque on resumption of rotation should not exceed 3.0 Nm.

Conveyor idler bearings are subjected to forces through the belt tension, the belt weight, the weight of the material loaded onto the belt and the weight of the rotating parts of the idler. The spacing of the idlers needs to be appropriate to the duty, having regard to the required bearing life. This life is the L_{10} fatigue life, which is defined as the life that 90% of bearings of a given type under given operating conditions are expected to exceed. This life can be calculated from a knowledge of the bearing's rated capacity (given in manufacturers' data) and the load on the bearing. For the bearing life calculations it is generally assumed that

Design of belt conveyors 2 – hardware

two-thirds of the load on the carrying idlers is borne by the centre idler. BS 8438[4.2] states that the minimum bearing life shall be 25 000 hours. However, British Coal, for example, required a minimum life of 50 000 hours. Clearly, the required life will affect both the size and type of bearing and idler selected and the spacing of the idlers.

CEMA[4.4] classifies idlers into four classes each coded by a letter (B, C, D and E) and a number that denotes the diameter of the idler roller in inches (4, 5, 6 and 7). Thus an idler might be classed as C4 or D6 and so on. Class B is considered to be suitable for light duty, classes C and D, medium duty and class E, for heavy duty. The different classes are suitable for different belt width ranges, which partly defines their required duties. The load ratings are related to the minimum L_{10} lives of the bearings in the idlers, with classes B and C having minimum L_{10} lives of 30 000 hours at 500 rev/min and classes D and E having L_{10} lives of 50 000 hours at 500 rev/min. CEMA provides a method of determining the idler type required, based on the load on the idler, with a load adjustment factor for the type of material conveyed, (in particular its density and the possibility of impact), an environmental factor (good, moderate, dirty) and a maintenance factor (good, fair, poor). The method allows for the hours per day the conveyor runs, and the belt speed and idler diameter.

We have seen in Chapter 3 that the minimum tension in the system should be such as to limit the sag between idlers to less than 2% (Equations 3.15 and 3.16). These same equations may be rearranged to allow the idler spacing to be calculated assuming the sag to be limited to 2% of idler spacing. We have seen from Equation 3.16 that for the carrying side

$$T_s = \frac{a_c \cdot g(q_B + q_G)}{8(h/a_c)} \qquad 4.1$$

where
  a_c is the carrying idler pitch (m)
  q_B is the mass of the belt (kg/m)
  q_G is the mass of the material conveyed (kg/m)
  h is the allowable sag (m)
  g is the acceleration due to gravity (m/s²)

Hence by rearranging and substituting

$$a = \frac{0.16 \cdot T}{g \cdot (q_B + q_G)} \qquad 4.2$$

This process of checking that the idler spacing is satisfactory for both the chosen bearing fatigue life and limiting the belt sag is part of the iterative process that is involved with conveyor design.

BS 8438 gives recommended idler pitches to suit various duties, which are measured by the density of the material conveyed and the belt width. Carrying idler pitches vary from 1600 mm for densities of 400 to 1200 kg/m^3 with belts below 600 mm wide, to 800 mm for densities from 2000 to 3800 kg/m^3 with the largest belts (1400 to 2000 mm). Return idler pitches of around 3 metres are usual.[4.2]

Idlers are produced in various standard diameters, ranging, for example from 101.6 to 168.3 mm,[4.2] or, as we have seen, 4 inch, 5 inch, 6 inch and 7 inch in the USA, designed to suit a range of duties. Concentricity of the idler barrels is important, as deviations in this factor will cause vertical displacement of the belt and lead to increased noise and vibration. Concentricity is affected by the quality of manufacture of the components and their assembly. The length of idlers has also been standardised and is selected to be in accordance with the width of the belt used.

In terms of design life, there may be a balance to be struck between the life of the bearings in the idlers and the life of the barrels if the environment is corrosive and abrasive. The UK coal mining industry always took the view that the life-limiting factor should be the wearing through of the barrels, on the grounds that it was better for this to happen than for the bearings to fail, possibly seizing and causing heating or fire. BS 8438 does not specify the type of bearing to be used in idlers; deep groove ball bearings are generally used, but in the USA taper roller bearings are sometimes specified, if first cost is not the deciding factor. Plain bearings have much higher internal friction than ball bearings with consequent increases in the power required to drive the conveyor. For a given shaft or outside diameter, taper roller bearings have a higher load capacity than ball bearings.

In the authors' experience, which relates to the UK, bearings often fail before the barrels, but by wear and seizure rather than fatigue, because the seal has failed and dirt and moisture have penetrated into the bearing. To help cope with dirt ingress the deep groove ball bearings used in some conveyor idlers have larger balls than normal bearings of the same outside diameter, polyamide cages and internal clearance higher than normal. Such bearings have been referred to as 'anti-seize' bearings. Considering the environment in which conveyors handling bulk materials work, it is clear that, as Ketelaar and Davidson[4.3] have pointed out, the design and performance of idler bearing seals is a key factor in the life of conveyor idlers. They are frequently exposed to abrasive minerals, sometimes to rain or pressure washing, possibly to extremes of temperature that may cause the seals to 'breathe' in water or dirt. Collapsed idler bearings are a significant cause of fires in underground coal mines. Much effort was devoted in the UK coal mining industry in the 1980s to making as many as possible of the components used in idlers fire resistant, including the lubricating grease used in the bearings in an attempt to limit the fire hazard on conveyors. However, there appears to be evidence that the phosphate ester-based fire-resistant grease used in the idlers does not lubricate as well as mineral oil-based grease and may lead to premature bearing failures.

4.2.3 Non-metallic idlers

Almost invariably in the past, steel has been the material of choice for idler barrels because of its durability and availability. However, steel corrodes and in some environments this can be a life-limiting factor. The authors recall trials of nylon-coated idlers in some UK mines where the mine water was extremely corrosive and idler barrels were destroyed in about eighteen months by corrosion-abrasion. We have seen from Chapter 3 that if the weight of the rotating parts of an idler could be reduced then the power requirements of the conveyor would also be reduced. Polymeric materials, which resist corrosion and are of low density, have been used for idler barrels. The design process is not straightforward, however. The cost of the raw material is important, as is the cost of processing and the ability to process it into the appropriate shape. The material needs to have the right stiffness to make a barrel, otherwise the wall thickness may be so large as to make the barrel too expensive. UV resistance is needed, together with wear resistance, although none of the common polymeric materials can match steel in the absence of corrosion. Some plastics, such as nylon, absorb large amounts of water and swell significantly, so are not suitable from the point of view of dimensional stability.

When all of these design factors are taken into account the choice is limited. High density polyethylene (HDPE) has been used for idler barrels and in recent times polypropylene incorporating inert fillers has been introduced with claims that it can last up to four times as long as HDPE. These materials have substantial advantages in terms of weight relative to steel and therefore reduce the rotating mass used in the calculation of power required to drive the conveyor. This weight saving is also valuable in terms of physical handling by personnel needing to install or replace idlers. They are also corrosion resistant, unlike steel, and in certain circumstances can provide greater wear life where a combination of corrosion and abrasion is present. A further advantage of these idlers is noise reduction, which can be an important factor. However, because of its lack of fire resistance and the fact that it is not inherently anti-static, it is not suitable for use underground in mines. (Chapter 8 discusses safety issues including fire resistance and static electricity in hazardous situations.)

4.3 Conveyor structure

The essential feature of conveyor structure is that it should provide a firm and stable basis for the mounting of pulleys and idlers, whilst allowing these items to be adjusted to allow the conveyor belt to run to its intended course without fouling either the structure itself or other objects.

In 1987, the former British Coal Corporation (BCC) issued a revision to its Specifications 539[4.5] and 540[4.6] to cover the design of the structure upon which was mounted the conveyor idlers and the conveyor belt. In many ways, this was an exercise to bring within the commercial control of this large mining company the designs of lightweight, easily assembled belt structure designs that were already being supplied by conveyor manufacturers and which were in widespread application in its mines and in the wider materials handling industry.

Belt structure effectively gives a fixed or planned pulley position in both horizontal and vertical planes to allow the conveyor to operate as designed, thus maintaining clearances between belt and surroundings. The BCC specified structure was also required to be fire resistant, and perhaps more importantly, to allow the easy clearance of any spillage of material away from rotating idlers. As mineral extraction and tunnel construction involve the changing of feed and delivery points at regular intervals leading to certain conveyors being frequently extended or reduced in length, the design of the structure had to facilitate change in length with maximum ease and minimum disruption to extraction operations.

Structure comprises stools and stringers. Stools are constructed from two vertical stanchions connected by a single horizontal cross tie. A plain steel square foot is attached to each stanchion leg. Brackets to mount the return idler spindles are mounted on each stanchion leg. The stringer is a tubular steel member that locates between mountings fitted to the top of each adjacent stanchion. Structure can either be placed on the ground, or in the case of tunnels or mine roadways, be suspended from the roof or wall. Top idlers are either mounted on fixed brackets at appropriate intervals on top of each stringer or as flexing units strung by end fixings between stringers (Fig. 4.4). Whilst the modern breed of long distance, high power, high speed conveyors might use much stronger and heavier structures, the basic design principles of the carrying structure remain largely unchanged.

In the surface mining industry, it is sometimes necessary to move the whole conveyor sideways in the horizontal plane with a minimum of dismantling. This is achieved by mounting the structure on a sub-base of rail track and sleeper bed. Sideways movement is achieved by progressively pushing laterally, sections of the rail track (Fig. 4.8).

4.4 Pulleys

The basic design parameters for the manufacture of conveyor pulleys are matters of straightforward mechanical and metallurgical engineering. Pulley design is fully covered in standards and manuals (see, for example, references 4.2, 4.4, 4.7, or 4.8). What matters to the conveyor designer in particular is that the pulleys should have the correct diameters and have the correct coefficient of friction for the transmission of power to the belt. The size of pulley is, of course, important in terms of the belt speed, but the stresses induced in the belt by its passing round the pulley also need to be considered. These stresses arise basically because the outside of the belt travels further than the side contacting the pulley and need to be limited to avoid damage to the belt. While in theory

4.8 Belt structure units mounted on rail track to allow lateral movement.

there could be an infinite number of pulley diameters, standardised ranges of pulley sizes have been introduced. ISO 3684[4.9] provides guidance on the choice of minimum pulley diameters and standards such as DIN 22101[4.10] and BS 8438[4.2] provide tables of recommended pulley minimum diameters appropriate to belt construction that vary with:

- the anticipated maximum belt tension
- the thickness of the belt
- the materials from which the belt carcase is made
- the percentage of the maximum allowable belt stress that is applied.

The diameters are largest for those applications such as drive pulleys where the tension is highest and smallest for bend pulleys where the change of direction is less than 90°. Recommended minimum pulley face widths for standard belt widths are also provided in the standards such as BS 8438.

The vast majority of pulleys have flat faces where they contact the belt. However in some instances so-called 'crowned' pulleys are used. The diameter at the centre of the face of a crowned pulley is larger than that at the outside edges. The extent of crowning is small. BS 8438[4.2] specifies that the pulley nominal diameter, i.e., the diameter in the centre of the pulley, shall not be more than 1% greater than the pulley edge diameter. There are various methods of crowning; it may be in the form of the arc of a circle or of straight-sided tapers that may or may not meet in the centre of the pulley. BS 8438 requires that a flat central portion shall not exceed 40% of the width of the pulley. Crowning is said to aid the tracking or training of a belt, i.e., may assist in getting the belt to run to its intended path, but it does introduce additional stresses into the belt and is therefore not favoured by belt manufacturers.

The coefficient of friction between belt and pulley is important for the transmission of power, the magnitude of the belt tension and the relationship between T_1 and T_2 (see tension calculation section in Chapter 3 Section 3.5). This friction coefficient can vary widely depending on the nature of the pulley surface, the type of belt and the conditions under which the conveyor is operating. Pulleys are generally made from steel but the surface of the pulley may be covered, or lagged, with rubber, polyurethane or ceramic to increase the coefficient of friction under certain conditions. The lagging may be grooved with a herringbone pattern. BS 8438 gives guidance figures for rubber belting contacting a bare steel pulley and pulleys lagged with the materials mentioned. Values of friction coefficient vary widely, from 0.05 for a bare steel pulley in wet, dirty conditions to 0.35 for a ceramic-lagged pulley in the same conditions, and to 0.45 for ceramic or rubber lagging in dry conditions. The UK coal mining industry has typically used a value of 0.25 for PVC covered conveyor belts running against bare steel pulleys. It generally avoided using lagging underground because of the danger of generating fires through friction of the belt against the lagging in the event of belt slip or stalling (see Chapter 8). Indeed, the UK Health and Safety Executive in its report 'Safe use of belt conveyors in mines'[4.11] states:

> Conveyor drives continuing to run while a belt is in a stalled condition is the most common cause of smoke emission. Frictional heat can be developed quickly and large volumes of acrid smoke and toxic gases given off from the heated belt. Drive drums lagged with fire resistant polymeric material give better adhesion than steel drums, but when slip occurs, smoke emission is made worse. With adequate design, installation and maintenance, the need for lagged drums is unnecessary and they should not be fitted.

4.5 Conveyor drives

As we have seen in Chapter 3, the arrangement of pulleys in conveyor drives can vary enormously in complexity from the single pulley to systems with several driving pulleys. Despite this variability, all drives must have at least one prime mover, which is usually an electric motor and a speed reducer to convert the motor shaft speed to the appropriate pulley shaft speed. Other components may be present to give control of the starting, running and stopping characteristics of the conveyor. Fayed and Skocir[4.12] have written extensively on the performance and characteristics of electric motors, speed reducers and other components used in conveyor drives, and it is not our purpose here to try to repeat what they have so ably produced. The reader is referred to their publication if he or she needs detailed information on these pieces of equipment. What we wish to do here is, by providing a little of the history of the development of conveyor drives, to highlight the essential features that have needed to be incorporated as conveyor duties have become more arduous and conveyor systems more complex.

As we relate in Chapter 2, the earliest conveyor belts were of relatively low strength, with the result that the loads that could be conveyed and the distances over which they could be moved were also low. Streets reports[4.1] that prior to 1905 the use of drives incorporating more than a single driving pulley was virtually unknown, principally because conveyors were short and of low power. He relates that in that year, the first tandem drive of which there is a record was introduced into the UK coal mining industry. Subsequently, substantial improvements in the manufacture of belts produced stronger, multiple plied belts, which although still made from cotton, could accept much higher power inputs, so that higher tonnages could be conveyed over longer distances. Streets notes that between 1919 and 1935 a range of machines was introduced into the UK coal mining industry that comprised between two and five driving pulleys. These machines were designed to increase the total arc of contact between the belt and the driving pulleys so that greater power could be transmitted to the belt. We have also seen in Chapter 3 that multi-pulley drives also allow substantial reductions in belt tensions, so that either a lower strength belt could be used or the length of the conveyor could be increased. The inventor of the tandem drive, Richard Sutcliffe, was concerned by the need for compactness in the underground environment, a feature which was facilitated by the use of multiple drive pulley arrangements.

It will be clear from the discussion on tensions in Chapter 3 that belt tension is sensitive to changes in the actual (as opposed to the theoretical) coefficient of friction between the belt and the driving pulley. If the coefficient of friction is lower than anticipated the tension can be substantially higher than calculated, and Streets[4.1] suggests that for single pulley drives, larger driving pulleys are needed to reduce the bending stress as the belt passes round the pulley to compensate for the increase in stress from the lower friction coefficient. Streets remarks that the benefits to be gained from the use of multi-pulley drives are much greater than those derived from increasing the pulley size on a single-pulley drive. He gives figures for the diameters of drive pulleys for various situations and weights of duck to illustrate the size reduction achievable with multi-pulley drives. For example, for a particular 8 ply belt he gives diameters of drive pulleys of 32 inches (813 mm) for a single pulley drive, 25 inches (635 mm) for a tandem drive and 20 inches (508 mm) for a three pulley drive.

We might argue that Streets' concern that the coefficient of friction between the belt and the pulley is overdone, since conservative figures should be chosen, and that he was seeking to endorse the use of multiple-pulley drives. However, operating conditions can vary, particularly underground and be wetter or dirtier than anticipated. With multi-pulley drives the arrangement of the pulleys is of great importance, not only from the point of view of the stresses imposed on the belt, as we have seen from the case history in Chapter 3, but also because the arrangement may result in one of the driving pulleys running against the carrying side of the belt. The effect of this may be that the coefficient of friction between the belt and the pulley may not be that assumed in the design calculations and indeed may vary as the composition or nature of the material being conveyed varies. Clearly this has implications for the tension in the belt and load sharing between drive pulleys. The Omega drive configuration shown in Fig. 3.9 has the advantage of letting the belt run on the clean side for both driving pulleys.

Streets appears not to favour the use of multiple motor drives. He argues that these types of drive were introduced to allow some differences in the peripheral speeds of the driving pulleys. However, he remarks that the results from large numbers of geared tandem drives, in which the pulleys must have the same peripheral speed, did not suggest a need to move to multi-motor drives. Variabilities in friction coefficients between the belt and the pulleys, and difficulties in the electrical connections also caused concern. However, having multiple motors must provide the designer of the drive with more options for the drive arrangement and Kirk[4.13] points out that an advantage of the two drive pulley system is that either two or three motors can be used. With three motors the drive can be run at 70% of rated power in the event of the failure of one of the motors or gearboxes. When the drive is operating in arduous conditions such as those underground and may be some distance from the mine exit, this can be a very substantial advantage. Glossop[4.14] indicates that, despite Streets' reservations, this two driving pulley configuration had become widely accepted in UK coal mines by the mid 1980s.

We have mentioned the fact that the prime mover for conveyors is most often an electric motor. Depending on the type and design of the motor, it may be able to supply up to 250% of full load torque on start up. The result of such high torques being applied rapidly to a conveyor, giving shock starts or uneven accelerations, may be the production of high tensions and, consequently, damage to the belt, the joints or other components, or the production of tension waves in the belt. It is generally recommended that starting torque be limited to 150% of full load torque to minimise these problems. Whilst an electric motor needs to reach full speed as quickly as possible to avoid overheating of the windings, a conveyor generally needs to start slowly and build up to speed gradually. Various devices may be interposed between the prime mover and the speed reducer to give a more gradual or 'soft' start. The two most common such devices are the fluid coupling and the wet clutch.

There are various types of fluid coupling, the simplest of which, the fixed fill fluid coupling, is used for low to medium power conveyors with relatively simple profiles. The fluid coupling has two opposing impellers that are enclosed within a casing containing fluid. One of the impellers and the casing are attached to the prime mover and the other impeller, termed the runner, to the output device, usually the gearbox. When the prime mover runs up to speed, the impeller and the casing attached to it also rotate, causing the fluid in the casing to flow outwards forming an annulus. There it meets the runner and imparts energy to it, causing it to rotate. The conveyor increases in speed gradually as the torque is transmitted from the input impeller to the runner through the fluid. Fixed fill volume fluid couplings have been used for many years, are of relatively simple construction and have proved to be reliable. However, the fixed fill fluid coupling has limitations and is not suitable if greater control of speed and torque is required.

Other types of fluid coupling that have been used in bulk handling for many years enable greater levels of control. In these couplings, known as variable fill couplings, the level of fluid in the casing containing the impeller and the runner can be varied, and along with it the speed and torque supplied to the drive pulley. These couplings are used for more complex installations and have the advantage of allowing the motor to start unloaded.

Clamp[4.15] cites two features that were just becoming available at the time of his writing (1998), which he felt conveyors needed to have. The first of these was speed control, which would allow the soft start and the slow build up to high speed for mineral clearance, together with two other lower speeds, one for the transport of men (man-riding – see Chapter 9) and a lower speed for belt inspections and repairs. The second feature was torque control. He noted that while the existing fluid couplings in general use limited the torque passing from the motor to the belt during start up, once the conveyor was up to speed and the fluid coupling full, the coupling could allow the motor to transmit up to 3 times full load torque to the belt in a hold-fast situation. He saw no reason why electronic control should not limit the torque in such a situation to 1.5 times full load torque. He also remarked on trials being conducted on equipment intended to provide these features.

It is worth mentioning here the existence of a device known as the Universal Control Drive (UCD),[4.16] which was produced in the mid-1980s and was intended to replace the fluid coupling. This device incorporated a wet clutch and a torque converter, and was designed to provide a soft start facility and variable speed. A wet clutch basically consists of two interleaved sets of clutch plates immersed in oil, which lubricates and cools the plates. The input plates are connected to the input shaft and are free to move axially. When the plates are brought together with the output plates, power and speed are transferred to the output plates through the oil film between them. In the UCD, output speed was controlled by an electro-hydraulic valve activated from an external source to suit the need of the system. The torque converter allowed applications such as conveyors, which require high starting torques, to be accommodated. The UCD was claimed to be able to provide variable speed, torque control and torque regulation which could limit the tension in belts. The UCD was originally produced for two purposes. Firstly, to control the speed of conveyors in order to provide a predetermined feed rate of material onto a main trunk conveyor. The activation of the electro-hydraulic valve was from a belt weigher, which as the name indicates is a device for measuring the amount of material passing along a conveyor. Its second purpose was to allow the speed of conveyors to be reduced so that men could ride on the conveyor safely (see Chapter 9).

The development of the UCD appears to have been curtailed, probably by the rapid decline in the size of the UK mining industry during the late 1980s and early 1990s. However, it seems that the trials referred to by Clamp were successful, for as control technology has progressed, a wide range of drive systems has become available that fulfil not only his but many other requirements and provide a degree of speed and torque control hitherto unknown in conveying technology. Using appropriate control systems variable fill fluid coupling can provide the facility for regulating load sharing between drive pulleys in multi-motor drives, controlled acceleration ramps, control of the torque transmitted during acceleration, regenerative load control and creep speed. In addition to these systems and controlled slip clutch systems referred to above, which can provide similar features, drives, known as variable frequency drives (VFDs), or variable-speed drives (VSDs) have been introduced. These drives use sophisticated electronics including frequency inverters to control the rotational speed of an electric motor by controlling the frequency of the electrical power supplied to it. They provide the functions that the UCD offered, and give:

- continuously variable speed control
- soft start facility and controlled rates of speed increase on start-up
- the ability to match belt speed to production rate to provide a constant feed rate
- the ability to run continuously at low speeds
- torque limiting facility to protect against overload.

The ways in which these VFD or VSD drives work is beyond the scope of this book, and indeed the competence of the authors to explain: it is to be expected that the development of this technology will continue to progress. The important point is that proven technology now exists to provide the facilities described. Each of the types of drive system has its advantages and disadvantages and with greater complexity comes greater potential for problems.

The motorised pulley is a type of drive that, like the VFD/VSD, eliminates the need for a device between the prime mover and the speed reducer. As the name suggests, the motor and the gearing are housed inside the pulley shell, which is oil filled. The gear pinion is attached to the motor rotor and engages with the gearbox, which in turn transmits torque to the pulley shell. As with other AC motors this device may be controlled in the same way as the VFD drive through the use of frequency inverters.

We have seen that the use of multiple pulley drives can reduce the tension levels in the belt. However, on very long conveyors, belt tensions can become very high and an extension of multiple pulley principle is to position additional drives at calculated positions along the length of the belt. These drives, referred to as booster drives, allow greater total installed power to be used in the system

Design of belt conveyors 2 – hardware

as well as enabling tensions to be reduced and lower strength, lower cost belts and supporting structure to be used. Further to this, of course, as we have said previously, the use of one long conveyor instead of several shorter conveyors eliminates multiple transfer points and drive, tail and tensioning units. The booster may be an additional short conveyor that transmits power to the main conveyor by friction with the main belt riding 'piggy back' on the booster belt. It is important for maximum efficiency that the speeds of the main and booster belts are accurately matched during start-up, acceleration and full speed running.

A second type of booster, known as the tripper booster because of its configuration, inserts additional drive pulleys, around which the belt must pass, into the system. The belt passes round the additional drive pulley, which of course reverses its direction, and then round a return pulley to return it to its original course. This arrangement results in an additional transfer point at each location where this type of booster is installed, but still eliminates the tail and tensioning units. As with the piggy-back booster it is important to match the performances of the main and booster drives during start, acceleration and full speed running so that they are not 'fighting' each other. Control of these boosters is achieved by mounting a tension sensor behind the return drum of the tripper booster. The main drive motor accelerates up to full speed on a programmed ramp and the tripper booster motor accelerates as the tension sensor senses an increase in tension. Similarly, if the tension sensor senses a decrease in tension the booster motor slows down. The response of the system is reported to be such that with the drive systems now available, excellent control of tensions can be achieved.

These systems are still developing, of course, and the technology is being pushed ever further by the use of boosters in conveyors that have horizontal curves. A booster is positioned just prior to the curve to reduce the overall belt tension so that edge tensions in the curve can be contained within acceptable limits. Where very complex arrangements of boosters are used then great care is needed in arranging the control system to allow for such factors as differences in loading conditions.

Whilst the sophisticated drive systems available today can provide speed control that includes retardation, in the event of a power failure or the mechanical failure of a key component, all such control may be lost. A loaded inclined conveyor where the head is higher than the tail end may under these circumstances tend to run backwards. Clearly this is dangerous and could result in material piling up at the tail end. To prevent this happening a conveyor may be fitted with some form of anti-runback device that operates in essence like a bicycle freewheel and allows the conveyor to move in one direction only. If, on the other hand, the head is lower than the tail, i.e., the conveyor is regenerative, it may need to be fitted with brakes to avoid the conveyor running away if the power or some key component fails.

4.6 Methods of tensioning belts

We have seen that the minimum tension in the system needs to be such as to avoid belt slip at the pulleys and to restrict the amount of sag between idlers. It is usual to include a device in the system that provides both a means of applying the correct slack side tension and of taking up any extension, or 'stretch' of the belt that has occurred under load. Such a piece of equipment is, not unnaturally, known as a take-up device. Conveyor belts, being in the main made of polymeric materials, suffer both temporary (elastic) and permanent (plastic) stretch as a result of the tensions applied to them. As we have seen, these tensions can vary during the operating cycle of the conveyor as it accelerates or decelerates and becomes loaded or unloaded. We therefore have elements of permanent stretch due to the belt tension and elastic stretch due to the variations in tension during the operating cycle. In addition, the take-up device will need to accommodate tolerances in the installation parameters, such as belt lengths. Different types of belt have different stretch characteristics. Steel cord-reinforced belts have very low stretch whereas fabric carcase belts have a total stretch of between 1.5 and 3.0% depending on their construction. This topic is covered in more detail in Chapters 5 and 7.

For very short conveyors (less than 80 m) take-up adjustment may be in the form of screws operating on the bearing housings of the tail pulley. With this type of take-up, there is no measurement of tension and the setting of the tension can be a rather hit and miss business. On longer conveyors the simplest type of take-up uses gravity to maintain a predetermined tension. A loaded pulley is installed on the slack side, with sufficient travel to provide the appropriate amount of take-up (Fig. 4.9). Clearly this arrangement requires sufficient space to be available beneath the conveyor. Where this is not the case, the gravity take-up may be modified so that it acts on the take-up pulley as shown in Fig. 4.10, in which a tower is used. With the arrangement shown in Fig. 4.10 the take-up pulley is mounted on a carriage on a track so that it can move as required under the influence of the weight applied and in response to changes in length in the system. Figure 4.11 shows a gravity tower system on an overland conveyor.

Gravity towers are widely used and eminently suitable for conveyors of fixed length and where there are no height constraints. However, if space is not available for a gravity-controlled system, as is generally the case in underground mines, and if the conveyor may vary in length, then an alternative method of maintaining tension is through a powered rope winch attached to the take-up pulley. The tension is controlled by a load cell that activates the winch. The tension may be applied by the use of a hydraulic motor driving the winch system. A continuously variable pump drives the hydraulic winch. The pump is preset to maintain a fixed pressure, which, since pressure is converted into tension by the winch system, ensures a constant tension during all running conditions of the conveyor.

Design of belt conveyors 2 – hardware   61

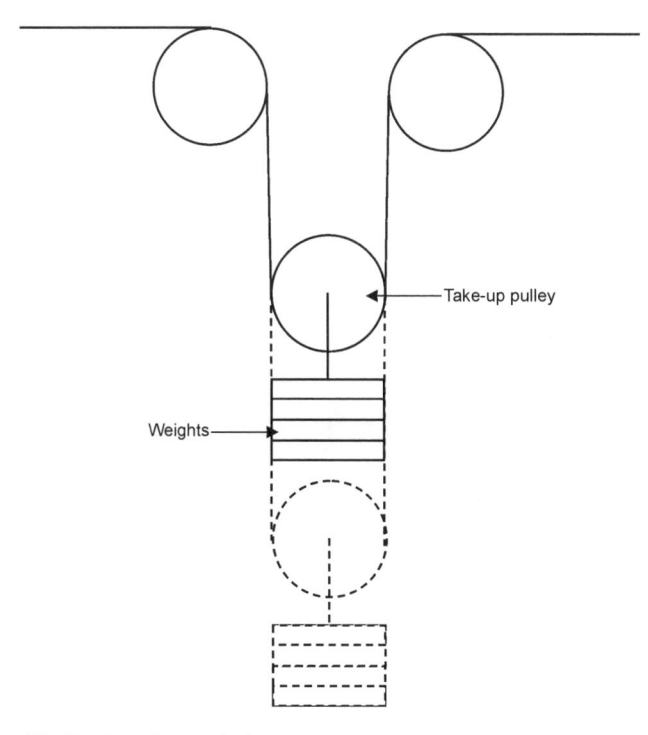

4.9 Gravity take-up device arrangements.

An alternative to the hydraulic winch is the use of the Variable Frequency Drive (VFD) arrangement. A combination of VFD and gearbox can be used in a winch configuration to provide a constant tension to the belt at all times, again by a rope system attached to the moving carriage in the tensioning arrangement. With this system different tensions can be applied for different loading and operational conditions. A brake is fitted to ensure tension is maintained during a power failure or emergency shutdown situation. Constant tension winch systems automatically adjust tensions for start-up, running and stopping, and as we remark above, may be used where space and height are not readily available for gravity systems.

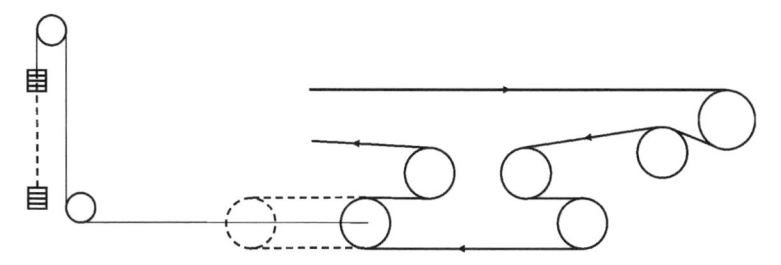

4.10 Diagrammatic representation of a gravity tower tensioning system.

62 Belt conveying of minerals

4.11 Gravity tower on an overland conveyor (by courtesy of ATH Resources Ltd).

Some conveyors in underground coal mines need to be either extended or retracted frequently to follow the progress of the coal workings, and for these types of installation the tensioning device is often combined with a belt storage device. The whole is referred to as the loop take-up. In a typical loop take-up the belt passes backwards and forwards over a series of pulleys, the distance between which can be varied to take in or release belt as required. If the conveyor needs to be extended, the tail end is pulled forward, a section of support structure is installed and the necessary amount of belt is paid out from the loop take-up. Modern loop take-up units can store up to 800 m of belt, allowing the conveyor to be extended up to 400 m without the need to fit additional lengths of belt.

ISO 3870 'Conveyor belts (fabric carcass), with length between pulley centres up to 300 m, for loose bulk materials – Adjustment of take-up devices'[4.17] provides some guidance.

4.7 Design of transfer points

On conveyor belt systems, it is often necessary to deliver conveyed material from one conveyor to another. This is usually as a result of a change in direction in the system flow. However, this transfer process can also result from other reasons. Delivery onto a conveyor has to be carefully designed when there is a significant increase in the speed of the receiving conveyor over that of the feed. Also, delivery systems have to be carefully engineered when the feed onto them is from a batch process rather than of a more regular flow as when one conveyor delivers onto a similar one.

The objective is quite simple. Capacity of the system has to be maintained at a maximum whilst spillage or blockages are minimised or eliminated. This has been a long-term problem with conveyor systems. In his 1979 paper, Watt[4.18] noted the twofold problem presented by belt conveyors. Firstly they were seen as the source of many delays to continuous production. Secondly, they were extremely demanding on manpower. He quotes a figure of 20 000 men being employed each day on 6000 belt conveyors in UK coal mines.

The reason for the high manpower requirement was that in the early days of conveyor operations in coal mines in the UK (up until mid-1970s), most transfer points were attended by an operator whose duties included cleaning spillage, and stopping operations if a blockage began to occur in the chute linking the two conveyors. Poor reliability in transfer point design frustrated the introduction of man-less transfer point operations and remote control and monitoring of conveyor systems.

An early study of transfer point design[4.19] concluded that the key design features of a successful transfer point were:

- continuous mineral transfer without blockages developing
- centralised delivery of feed onto the receiving belt to prevent mis-tracking and spillage
- control of mineral flow to prevent damage to the receiving belt
- adequate belt cleaning of the delivering belt to prevent carry-back
- mineral reclamation from belt cleaners.

To this list might be added the need to combat dust make caused during the flow of material through the chute and by the impact of the falling mineral on the receiving belt. So, in practice, complete enclosure of the chute to contain dust is not unusual (Fig. 4.12) and water sprays are strategically sited to dampen down dust.

The design study focused on chute design as this was seen as the key to efficient transfer between conveyors. It noted that gravity chutes had three associated problems that made it difficult to achieve the idealised design features listed above. These were:

- blockages from large lumps or objects
- blockages or flow disruption resulting from fines deposits
- poor centralisation of flow delivery onto the receiving belt.

4.7.1 Blockages from large lumps or objects

With regard to large lumps, it was recommended that lumps running on belt systems should not exceed 75% of the clearance in the chute, or have a

4.12 Enclosed chute at transfer point (by courtesy of ATH Resources Ltd).

maximum dimension greater than 250 mm. This requirement led to the installation of in-system product sizers very early in the materials handling cycle.

As with nearly every aspect of belt conveyor design, the CEMA publication[4.20] has a section devoted to transfer from one belt to another. It notes the importance of the inclination of the chute delivery section relative to the load being delivered. There is a compromise that has to be made between a steeply inclined chute that allows fine and wet material to slide down without sticking, and a low angle delivery to control lump material delivery speeds.

CEMA notes that a chute delivery should be no greater than two-thirds the width of the receiving belt, but also requires the inside dimension of the chute

> to be at least two-and-a-half to three times the largest dimension of uniformly sized lumps, when they represent a considerable percentage of the material flow. Where lumps and fines are mixed, the inside width of the chute may be as small as two times maximum lump size.

This does conflict somewhat with the 75% clearance requirement defined by Astle.[4.19]

4.7.2 Blockages or flow disruption resulting from fines deposits

Fines deposits usually become significant when wet. Increased dust suppression sprays, whilst dealing with airborne dust from conveyors, have added to the wet fines problem, as has the residue released by belt cleaners. However, once fines have stuck inside the delivery chute they tend to dry out and harden, building up a deposit reducing clearances further.

4.7.3 Poor centralisation of flow delivery onto the receiving belt

The centralisation of flow is also important. If the receiving belt is loaded to one side, the off-centre weight draws the mineral load towards the centre of the trough but it also drags the conveyor belt with it. This can ultimately lead to mineral on the receiving belt spilling over the edge of that belt.

Flow centralisation can also be managed by the fitting of skirt boards along the sides of the receiving belt that centralise the delivered flow in the first few metres of its run on the receiving belt.

The problems associated with transferring material onto higher speed belts is discussed by Fayed and Skocir.[4.12] They report the damage that can be caused to the receiving belt covers from material 'that billows out and surges in all directions' when loaded onto a faster moving belt. They note that 'material being loaded should be travelling at a speed comparable to that of the belt it will be loaded on to'.

The steel cord conveyor installed in the South Drift at Gascoigne Wood mine in the UK Selby complex that is described in Chapter 11 was, in its day, one of the fastest conveyor belts working, conveying mineral at speeds of up to 8.4 m/sec. It was recognised that delivery of mineral onto a belt travelling at this elevated speed from a feed hopper would require a new approach. Loading onto this conveyor required the installation of an in-line variable speed conveyor that would accelerate the load to a speed at which it could be safely delivered onto the main conveyor without damaging the conveyor belt or causing spillage.

A research project conducted by the British Coal Corporation (BCC) to establish the most appropriate speed for the accelerator conveyor was conducted in a full size surface facility and concluded with many caveats that a feed conveyor delivery speed of 3.5 m/sec was required in this installation.[4.21]

Transfer of mineral from a batch production process, such as a shuttle car supporting a continuous miner, onto a belt conveying system requires careful design of the loading facility to ensure an even flow of suitably sized material onto the belt conveyor. In these circumstances, mobile feeder–breaker units are employed. These comprise a receiving section bunker to take the full load of a shuttle car, a sizer unit to break lumps and a slow-speed scraper chain loader that provides a steady consistent feed onto the belt conveyor. This unit can also be linked to the tension end of the belt conveyor, both tension end and feeder breaker being advanced as the mining area advances.

One further feature of the design of the transfer point that is worthy of brief comment is the need to protect the receiving belt and its surrounding structure from damage caused by the impact of material being delivered through the chute. Various devices have been designed to give a 'soft' landing to material, the most common being impact idlers (see Section 4.2.1) and slider beds. Slider beds are a series of bars made from polymeric material that run parallel to the major axis of the belt. These bars are in contact with the underside of the receiving belt and have a low coefficient of friction allowing the belt to run freely over them. When material is discharged onto the belt, the impact is transferred through the belt onto these bars, which have a degree of flexibility thus reducing the damage to the receiving belt.

4.8 Belt cleaning

All belt conveyors that are involved in the transport of minerals are prone to a residue of (often wet) material sticking to the surface of the belt after delivery discharge. Left unchecked, this material would remain stuck to the belt surface during its return run, falling from the belt randomly or being dislodged by the return rollers, in both cases causing spillage under the belt. This can be a particular problem with flammable minerals as the spillage can dry to become a fire hazard in close proximity to moving conveyor parts such as return idlers. It is normal to deal with this phenomenon by installing belt cleaning devices close to the head end delivery point. It is also clear that the mineral removed by the cleaning device has to be managed in such a way that it does not represent a subsequent hazard or even an untidy mess.

There are very many designs of belt cleaner and suppliers or manufacturers offering different approaches to the removal of mineral. Device design generally falls into one of the following headings:[4.22]

- scraping with rigid blades
- scraping with spring mounted or movable blades
- brushing with rotating brushes
- high pressure water or air sprays
- belt washing bath
- vibrating 'belt beating' devices.

Of these, the first three are the most commonly found. Figure 4.13 illustrates examples of belt cleaners displayed for exhibition purposes. It should be noted that in operation, the belt cleaning devices are often fully enclosed and therefore are hidden from view.

The most commonly employed designs are scraper blades with cleaning tips made from rubber, plastic, ceramics, steel or tungsten carbide.[4.22] Key design factors that should be addressed when considering belt cleaner performance include:

- the achievement of constant contact pressure between cleaner and conveyor surface
- the minimising of blade wear
- the condition of the conveyor belt surface
- the type of material sticking to the belt
- belt speed
- system vibration.

Design of belt conveyors 2 – hardware

4.13 Belt scrapers.

Most cleaning devices are found to be effective in the short term, but wear can quickly reduce this effectiveness.

Scraper blades can be used in three attitudes relative to the belt surface. These are either inclination towards the direction of belt movement (the paint scraper principle), or perpendicular to the belt surface (the bar principle) or inclined back from the direction of belt travel (the negative angle principle). The first of these is said to be the most effective.[4.22] Scrapers with rigid single blades spanning the width of the belt are prone to wear more quickly and lose contact with some of the belt surface allowing material to remain stuck to the belt surface. Modern higher efficiency scraper cleaners are designed in a split line layout with individual spring loaded elements in a line pressed against the belt surface. Single or double rows of scraper have been applied to improve effectiveness and often pre-scraper and main scraper groups or single units have been installed at different positions along the belt adjacent to the main delivery pulley. A study mounted by British Coal's Mining Research and Development Establishment[4.19] published in 1982 examined the performance of cleaning devices. The research recommended a multi-stage approach to belt cleaning. The Stage 1 or pre-scrapers were two rows of plastic-bladed belt cleaners to remove the coarse material. Stage 2 was a single hard-bladed (stainless steel) scraper to remove the fine slime. The author of this study recommended a relatively narrow contact strip across the centre of the belt. He also noted that steel blades should not be used with rubber or nitrile covered belts. The final or Stage 3 cleaner was a squeegee roller to remove any excess water. The report concluded that cleaning based on Stages 1 and 2 alone may be adequate dependent on conditions.

Scraper positioning relative to the main drive pulley is influenced by the clearance arrangement for the debris released by the scrapers. Whilst most effective cleaning can be achieved by scrapers placed on the drive pulley itself shortly after material is delivered from the belt, there is always a concern that a secondary flow of sticky material into the delivery chute may in fact help to promote blockages in the chute, reducing throughput. Scraper or screw type spillage conveyors have been recommended to feed the product of belt cleaning back on to the conveyor.

A United States Bureau of Mines (USBM) study conducted in 1989[4.23] concluded that an optimum pressure between conveyor and scraper exists that, when exceeded, leads to no improvement in materials removal efficiency but did lead to an increase in scraper blade wear. Non-uniform wear caused by uneven belt surface wear, belt joints and even company logos and other information recessed into belt covers allow material to pass between scraper and belt.

There is one further issue to mention about belt cleaning devices based on the scraper principle. Conveyor belting used in UK underground mines is constructed from materials with appropriate fire and electrical resistance properties. The application of a non-metallic belt scraper in intimate contact with the belt surface represents a situation in which an electrical charge can build on the cleaning blade. British Coal took the view that the material properties of any non-metallic scraper blade had to be compatible with those of the belting along which it had contact such that no dangerous discharges were possible.

A procedure that minimises the amount of material that may fall from a belt due to contact with return idlers is turning the belt over, so that the clean side of the belt comes into contact with the return idlers. This is not a practice that has been used in UK coal mines because of fears about overstressing of the belt edges. However, DIN 22101 provides guidance on the design of three types of turnover, those where there are no guiding or supporting rollers in the turnover, those where there is a vertical pair of rollers at the centre of the turnover and those where the belt is guided through the turnover by support rollers. The standard gives values for the lengths of the turnovers in terms of multiples of the belt width for cotton fabric carcase, synthetic fibre carcase and steel cord carcase belts. CEMA[4.4] also provides some guidance, but warns that turning the belt over does not obviate the need for conventional belt cleaners. Lemmon[4.24] states that existing methods of designing belt turnovers are based on twisting stresses and ignore bending stresses. He provides a calculation method for one type of turnover, the flat helix turnover, in which the belt is guided through its turnover by a central pair of rollers plus possibly two other pairs set at quarter distances.

4.9 High angle conveyors

The vast majority of conveyors moving bulk materials use smooth-surfaced belting. Because of this, the angle which the conveyor can adopt relative to the horizontal is limited by the tendency of the material to slide down the belt or to move internally relative to itself. Special conveyors have been developed to overcome these problems and to enable materials to be moved at angles up to and beyond the vertical. The developments include belts that have cleats or steps attached to the surface, belts that can be formed into tubes and conveyors that sandwich the material being conveyed between two belts. CEMA$^{4.4}$ deals comprehensively with this subject and the reader is referred to that publication if he or she needs further information on detailed designs and capabilities.

4.10 References

4.1 Streets H *Sutcliffe's Manual of Belt Conveying* W & R Chambers Ltd 1956.
4.2 BS 8438:2004 'Troughed belt conveyors – Specification'.
4.3 Ketelaar J P I and Davidson P J 'Improving the efficiency of conveyors used for the transport of minerals in underground and surface mines' *Mining Technology* Vol. 77 No. 883 March 1995.
4.4 Conveyor Equipment Manufacturers' Association (CEMA) *Belt Conveyors for Materials Handling* 6th edition 2005.
4.5 Flexing Troughed Belt conveyors (Structure and Idlers) British Coal Corporation Specification 539:1987.
4.6 Fixed Troughed Belt conveyors (Structure and Idlers) British Coal Corporation Specification 540:1987.
4.7 ISO 1536:1975 'Continuous mechanical handling equipment for loose bulk materials – troughed belt conveyors (other than portable conveyors) – Belt pulleys'.
4.8 ISO 1816:1975 'Continuous mechanical handling equipment for loose bulk materials and unit loads – Belt conveyors – Basic characteristics of motorised pulleys'.
4.9 ISO 3684:1990 'Conveyor belts – Determination of minimum pulley diameters'.
4.10 DIN 22101 'Continuous conveyors – Belt conveyors for loose bulk materials – Basics for calculation and dimensioning'.
4.11 UK Health and Safety Executive *Safe Use of Belt Conveyors in Mines – Topic Report*, HSE Books, HMSO 1993.
4.12 Fayed M E and Skocir T S *Mechanical Conveyors – Selection and Operation* Technomic Publishing Company Inc. Pennsylvania USA 1997.
4.13 Kirk A 'The installation of Grimethorpe high horsepower drift conveyors' *The Mining Engineer* November 1988 p 182.
4.14 Glossop M R 'Conveying higher tonnages' *The Mining Engineer* December 1984 p. 305.
4.15 Clamp G 'Transport safety in RJB' *Proceedings of Institution of Mining Engineers Symposium on Developments in Underground Transport* 1998.
4.16 Howcroft W, Bower L R and Wright G 'Universal Control Drives and their application to the mining industry' *The Mining Engineer* August 1986 No 299.
4.17 ISO 3870:1976 'Conveyor belts (fabric carcass), with length between pulley centres up to 300 m, for loose bulk materials – Adjustment of take-up devices'.
4.18 Watt R G 'Mining transport – short and long term considerations' *Mining Engineer* April 1979 No 211.
4.19 Astle R 'Belt Conveyor Transfer Point Design for Underground Applications' British Coal Corporation Mining Research and Development Establishment Report 82/62 November 1982.
4.20 Conveyor Equipment Manufacturers' Association (CEMA) *Belt Conveyors for Materials Handling* 5th edition 1997.

4.21 Waring B 'Testing the transfer of ROM material onto high speed conveyors used in the Selby complex' British Coal Corporation Mining Research and Development Establishment Report 83/40 November 1983.
4.22 Hell E 'Conveyor belt cleaning' *Quarry Management* October 2006.
4.23 Rhoades C A, Hebble T L and Grannes S G 'Basic Parameters of Conveyor Belt Cleaning' USBM Report of Investigations 9221, 1989.
4.24 Lemmon R 'Local stresses in belt turnovers' *Bulk Materials Handling* IV 2002.

5
Belt constructions

5.1 Introduction

The previous chapters on conveyor design have shown that there are basically two elements to the conveyor belt itself – the part that is in contact with, and carries, the bulk material, and the part that transmits and withstands the tensions imposed on it. The vast majority of belts are constructed so that the tension-transmitting element forms the central part, termed the carcase, of the belt, which is overlaid on both sides by covers that are made of materials with appropriate friction and wear properties. However, one construction, the cable belt, separates out the two elements by making the tension-carrying element into wire ropes, onto which the element that carries the bulk material is mounted, but to which it is not physically attached. For those types of belt that have carcases, the carcases may be of various textiles or consist of steel cords. We shall consider all of these varieties separately.

5.2 Textile carcase belts

Textile carcases are woven from a variety of materials and are of two basic constructions, plied and solid woven. Plied belts are by far the most common type of belt used in bulk materials handling. In this type, layers of woven fabric (plies) are built up with interleaved layers of rubber, which are bonded together under heat and pressure to form the belt (Fig. 5.1). In special circumstances, such as when the belt needs to be fire resistant, the plies are impregnated with polyvinyl chloride (PVC) rather than having rubber interlayers. ISO 14890:2003[5.1] defines 'monoply', 'duoply' and 'multiply' belts. The first of these has a carcase consisting of a single layer of fabric, the second two layers and the third more than two layers. The weft or transverse threads of solid woven belts are arranged in layers that could be considered equivalent to plies, but these are mechanically locked together by longitudinal warp threads that pass through the carcase thickness. Figure 5.2 shows a simple solid woven construction with two layers of weft threads and four binder warp threads

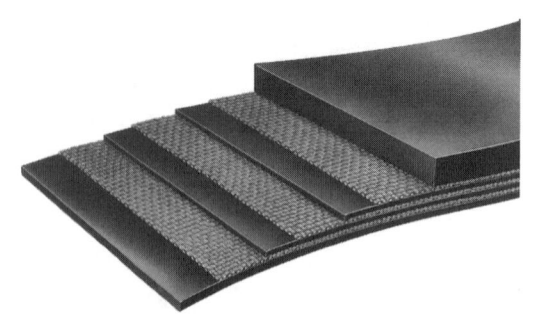

5.1 Plied belt construction (by kind permission of Fenner Dunlop Conveyor Belting Europe).

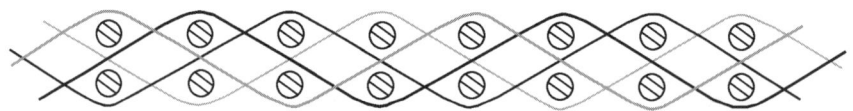

5.2 Simple solid woven construction (by kind permission of Fenner Dunlop Conveyor Belting Europe).

locking the two weft (ply) layers together. A three-dimensional view of a solid woven construction is shown in Fig. 5.3. Because of their construction, it is possible for plied belts to suffer ply separation, which is termed delamination, but with solid woven belts this cannot occur.

Solid woven belts came to prominence in response to the drive by the UK National Coal Board for fire-resistant belts following the disastrous mine fire at Cresswell Colliery in 1950[5.2] when eighty men were killed as a result of a belt conveyor fire. At that time it was not possible to make plied belts that were both satisfactorily fire resistant and that were of adequate quality for use underground. Fire-proofing of the carcases with chemicals still left the highly flammable rubber inter plies and covers to spread a fire. Replacement of the rubber by PVC, which is naturally fire resistant, gave quality problems that resulted in short operational lives. However, these quality problems were overcome and the

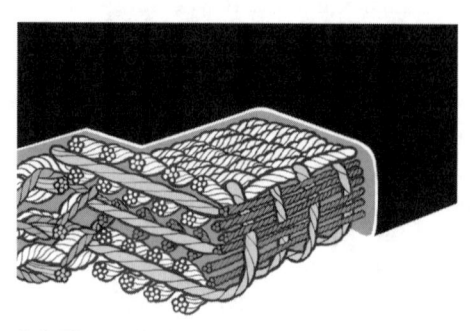

5.3 Three-dimensional view of solid woven belt construction (by kind permission of Fenner Dunlop Conveyor Belting Europe).

solid woven design impregnated with PVC eventually met operational and fire-resistance requirements (for more information see Chapter 8).

The design of textile carcase belts is not a simple matter and a great deal of technical development has gone into the design and production of belts since the early days of conveyors when canvas fabrics were used. The belt needs to have appropriate properties in both the longitudinal, or warp, direction and the transverse, or weft, direction. Some of these properties are achieved through the choice of materials from which the carcase is made and some through the way in which the materials of the carcase are put together. Yet others arise from the choice of cover compounds. Considering firstly the carcase, in the warp direction, the main property required is obviously strength, which arises from the number of warp yarns and for plied belts, the number of plies used, but flexibility and stretch are also important. In the weft direction, troughability, i.e., the ability to conform properly to the shape of the idler sets so that the belt contacts the idlers, is important, as is the ability to retain the mechanical fasteners that may be used to join lengths of belt together. Resistance to impact damage and tearing are features that can be built into the belt carcase, features that can be of great importance operationally. To this end belt constructions sometimes include a 'breaker', which may be an additional open mesh fabric, cord fabric or cord layer, positioned between the carcase and the cover to protect the carcase.

As mentioned above, it is not only the materials from which a belt is made, but also the way in which the materials are combined that is important in giving the finished belt the properties needed for a given application. Yarns of the same or different types are combined to provide the appropriate properties, before the weaving process. Tension is applied to the yarns during weaving to provide uniformity across the belt width and thus avoid problems such as differential edge tensions that might lead to tracking problems. BS 8438[5.3] (for example) mentions cotton, polyamide, cotton/polyamide, cotton/polyester, polyester and Rayon as possible materials for the load carrying warp threads of textile carcase conveyor belts. ISO 14890[5.1] designates the various types of yarn that might be used in textile carcases by code letters, as follows:

B = Cotton, Z = Staple Rayon, R = Rayon, P = Polyamide, E = Polyester, D = Aramid, G = Glass

In plied belts, polyester (Terylene) is the most common warp thread with polyamide (Nylon) being used in the weft. In solid woven belts the warp threads are usually either polyester or polyamide, as are the weft threads that provide the transverse strength needed in the belt. With some designs of solid woven belts, additional cotton warp yarns are included on the top and bottom of the carcase to provide a 'pile' that protects the load-bearing yarns from impact and assists in cover adhesion and aids fire resistance. Figure 5.4 shows a solid woven construction with protective surface yarns (piles). Polyamide has a lower modulus

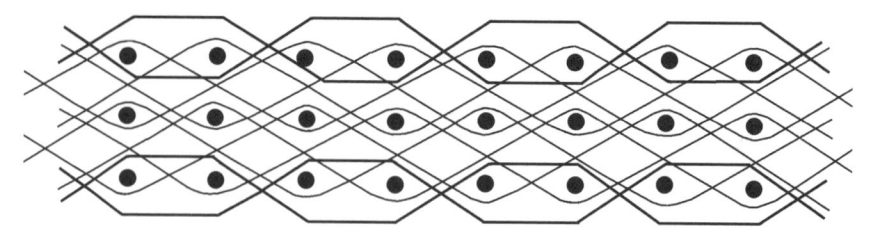

5.4 Solid woven belt with protective surface yarns (by kind permission of Fenner Dunlop Conveyor Belting Europe).

than polyester. This produces a belt with better flexibility that is suitable for use with smaller pulleys. It also has greater resistance to dynamic stresses, such as might be imposed by start-up or impacts, and greater resistance to abuse. The lower modulus also results in a belt with greater stretch in service, which needs to be considered when designing the take-up device. The lower stretch inherent in belts with carcases based on polyester is accompanied by rather lower flexibility than with polyamide carcases, so careful consideration of these factors is needed when designing the conveyor system. Solid woven belts are produced to given widths with the weft threads forming closed edges to the belt (selvedges). Multiplied belts, on the other hand, are generally made in 'slabs' and cut to width (cut edge belting). The weaving process for a solid woven belt is shown in Fig. 5.5.

5.5 Weaving of a solid woven belt (by kind permission of Fenner Dunlop Conveyor Belting Europe).

5.6 Impregnation of a solid woven belt with PVC by dipping (by kind permission of Fenner Dunlop Conveyor Belting Europe).

Following the weaving process the solid woven belt carcase is impregnated, either with PVC or with rubber (Fig. 5.6). For the most highly fire-resistant belts, impregnation is carried out with a PVC plastisol, which is specially formulated to impart the appropriate properties to the finished belt. The formulation may include plasticisers, stabilisers, fire retardants and other additives. Since PVC is itself inherently fire resistant, while the carcase is not, it is vital that in the impregnation process the PVC reaches right to the centre of the carcase. The properties of the PVC are important not only because they affect the fire resistance of the finished belt, but also because they can affect the physical attributes of the belt such as troughability and fastener holding.

The final step in the manufacturing process for PVC impregnated belts is the application of the covers. PVC covers are applied generally by dip coating but pressing may also be used. One of the problems that was found with PVC belts in the early days of their production was that the coefficient of friction of the covers was rather lower than that of rubber covers. This is important in two ways. Firstly, power is transmitted from the drive pulley to the belt by friction and the choice of an appropriate coefficient of friction for the design calculations is vital (see Chapter 3 Section 3.5). Secondly, belts with PVC covers cannot be used on such steep gradients as belts with rubber covers, particularly under wet conditions, because load slip can occur. As technology has developed, methods have been found to bond rubber covers over the top of the PVC covers. Synthetic rubbers such as neoprene (polychloroprene) and nitrile (polyacylonitrile) rubber are used and have enabled conveying gradients to be increased from 16° for PVC covers to 22° or more for belts with synthetic rubber covers. The synthetic rubber covers also confer additional abrasion and impact resistance. As an alternative to PVC, solid woven belts may be produced with rubber impregnation and rubber covers. Under a combination of heat and pressure the rubber covers are impregnated into the polyester-based carcase to become an integral part of the belt. With appropriate rubber compounding, these belts can offer the advantages of solid woven construction that gives a combination of flexibility that allows the use of smaller pulleys, low stretch (less than 2%) and a sufficiently high degree of fire resistance to meet all but the most stringent standards.

During the assembly of plied belts great care is required to avoid contamination or the entrapment of air between the layers of fabric and rubber as the belt is laid up, as these may lead to delamination of the belt in operation. Individual plies may have longitudinal or transverse joints in them and care is again needed to ensure that the joints butt up to each other without either gaps or overlaps. ISO 14890[5.1] specifies the number of longitudinal and transverse joints that are permitted in the outer plies and the inner plies for given lengths of belting and the spacing between joints in adjacent plies. Naturally, the standard does not allow either transverse or longitudinal joints in monoply or solid woven belting. Following the laying up process, the covers are applied to the belt and the whole assembly is vulcanised to fuse it together as an integral whole under heat and pressure. The covers perform the important function of protecting the carcase against abrasion and impact. In years gone by belts were made almost entirely from natural fibres and it was important to seal the edges of the belt to prevent the ingress of moisture and attack of the belt carcase by mildew. The belt carcase was therefore totally enclosed by the covers that extended round the edges of the belt. As we have seen, some belt types still contain cotton and thus need to be protected against moisture ingress. Most types, however, have carcases consisting entirely of synthetic fibres that are immune to mildew attack, and this has made possible the production of ply belting that is cut to size and that is put into service without sealing the edges. The vulcanising process may be done using large flat vulcanising presses or by the 'Rotacure' process in which a large diameter heated drum acts as one platen of the press and a moving steel band as the other. The Rotacure process avoids the periodic opening and closing process needed with a conventional flat press.

Scrupulous cleanliness is essential during the application of the covers to avoid blistering of the covers in operation. Similarly it is important to avoid the entrapment of air, which can also cause blistering. Such blistering can cause problems in the cleaning of the belt and can cause vibrations to be generated and transmitted to the supporting structure as the belt passes through the drive. Minor blistering can be repaired by puncturing the blisters and injecting an adhesive cement to reattach the blistered area. Rubber covers of belts generally have good abrasion resistance, but may be formulated to have special properties such as higher resilience, or higher abrasion resistance. Butyl (polybutylene) rubber is used for heat resistance, EPDM (ethylene-propylene diene monomer) for heat and ozone resistance, nitrile and neoprene for oil or, as indicated above, fire resistance. ISO 14890[5.1] refers to three grades, H, D and L, which are defined in ISO 10247[5.4] based on their tensile strength, elongation at break (measured according to ISO 37[5.5]) and abrasion resistance (ISO 4649 Method A[5.6]). Whilst PVC-impregnated solid woven belts are usually produced with either one- or two-millimetre PVC covers or with one-millimetre PVC plus two-millimetre rubber covers, the rubber impregnated solid woven belts may have up to eight-millimetre covers on the carrying side. Covers on the side of the belt carrying the bulk material are generally substantially thicker than those of the drive side that contacts the pulleys and may incorporate wear indicators (see below) if conditions are particularly severe, such as in the metalliferous mining sector. Cover thicknesses on ply belts may be up to 25 mm on the carrying side and up to 13 mm on the drive side.

A feature that may be incorporated into belts is the wear indicator. This may be in the form of a material that is a different colour from the belt cover and that is incorporated into the cover to show, by the change of colour that occurs when the material is exposed, when a certain depth of the cover has been worn away.

It was noted in Chapter 3 that belt widths have been standardised by national and international organisations. Belt strengths have also been standardised and are measured in newtons per millimetre width, kilonewtons per metre width, which is numerically the same, or, in the USA, pounds per inch width. This measurement defines the minimum breaking strength of a belt when pulled in the longitudinal direction. Further information on how this measurement is made is given in Chapter 7. BS 6593[5.7] notes that belt strengths run from 200 N/mm for 2-ply to 2000 N/mm for 6-ply. Solid woven belts are available from 400 N/mm up to 3150 N/mm. In many parts of the world, belts are still referred to by their old Imperial designations. Under this system the tensile strength in lb/in was divided by one thousand to give the belt designation. Thus a belt that had a minimum tensile strength of 8000 lb/in (equivalent to 1400 N/mm) would be known as a Type 8 belt.

As explained in Chapter 3 the tension at which a belt works on a conveyor, the running tension, is related to the tensile strength of the belt by the 'service factor' or 'factor of safety'. For textile carcase belts this has in the past traditionally been ten, so that the maximum running tension is one tenth of the ultimate tensile strength of the belt. The origins of this figure are not entirely clear, although Gilbert[5.8] remarks that years of experience have shown that this ratio is needed to optimise belt life and performance. He further remarks that the figure of 10:1 is not needed for pure tensile strength reasons, but because the associated belt characteristics meet operational requirements. However, carcase designs and the materials used in them are evolving all the time and some belt manufacturers may advise that lower service factors can be used in some applications.

Textile carcase belts are designated according to ISO 14890[5.1] by reference to:

- the width in millimetres
- fibre type in the warp and weft directions
- the tensile strength
- the number of plies
- the thickness of the covers in millimetres
- the cover classification and
- the safety category according to EN 12882[5.9] (see Chapter 8).

It is strange that this last requirement should appear in the ISO standard because the EN standard is only applicable in the European Union: this appears to be an oversight on the part of the standards writers.

Thus a belt designated

1000 EB PB 800 5 4 2 H 1

has a width of 1000 mm, warp yarns of polyester/cotton, weft yarns of nylon/cotton, a tensile strength of 800 N/mm, five plies, a top cover of four millimetres, a bottom cover of two millimetres, covers of type H and a safety category of 1. These designations are usually embossed into the carrying cover of the belt by 'branding'. The standard also requires the marking of the belt to include reference to the standard, the name of the manufacturer and the date of manufacture.

5.3 Steel cord belts

Because of their construction, steel cord belts have high strength ratings and can be made in tensile strengths that cannot be achieved with textile belts. This can enable extremely long conveyor runs to be installed. The strength elements of steel cord belts are stranded wire ropes which are laid longitudinally as a single layer and embedded in rubber to form the core of the belt, over which the covers are applied (Fig. 5.7). Depending on the application for which the belt is intended, the individual strands of the wire ropes or cords may be protected by galvanising or brass coating to combat corrosion that might arise if the belt cover is penetrated and water enters. The whole construction is vulcanised together under heat and pressure.

One area of weakness of this type of belt can be its susceptibility to damage.

Belt constructions 79

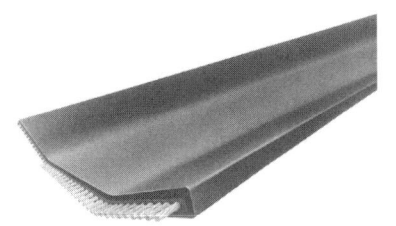

5.7 Steel cord belt (by kind permission of Fenner Dunlop Conveyor Belting Europe).

If a foreign object penetrates the belt cover, water can gain access to the steel cords and corrosion of the cords can seriously weaken the belt. If an object completely penetrates the belt, it can rip for great lengths or even for the whole length of the belt unless measures are taken to prevent this. Belts are therefore often made with the incorporation of a transverse reinforcement between the cords and the covers. This reinforcement may be a 'breaker', usually a woven fabric, or a 'weft', which is made of steel wires. The breaker or the weft may be incorporated both above and below the steel cords or either above or below them. For steel cord belts a breaker is defined as a transverse layer lying at least one millimetre and up to three millimetres from the cords of the belt core; it is considered to be part of the cover of the belt. A weft is defined as lying not more than one millimetre from the cords and is considered to be part of the core. Figures 5.8(a) and (b) illustrate these arrangements. The transverse reinforcement is designed, obviously, to provide transverse strength, but also to equalise stresses across the belt, to assist in preventing a foreign object penetrating through the belt and to dislodge it if it does penetrate. Much development effort has been devoted to the detection of rips in steel cord belts. This subject will be discussed further in Chapter 10.

The strength of the belt is defined by the properties of the steel cords, their diameter and their pitch, i.e., the distance between adjacent cords. Clearly vast numbers of combinations of cord diameter and pitch are possible to give a very large range of possible belt strengths and properties. The International Standards

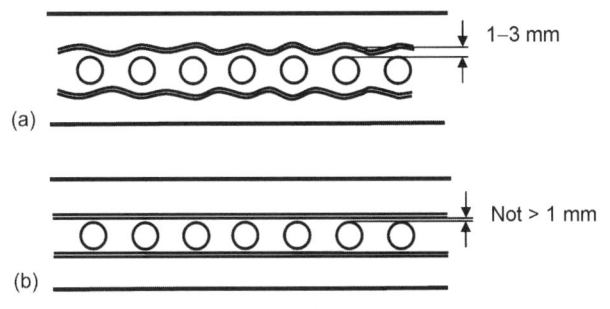

5.8 (a) Steel cord belt with breaker above and below cords, (b) steel cord belt with weft above and below cords.

Organisation (ISO) has therefore defined design, dimension and mechanical requirements for steel cord belts in ISO 15236-1:2005[5.10] and preferred belt types in ISO 15236-2:2004.[5.11]

Maximum running tensions for steel cord belt are higher relative to the ultimate tensile strength of the belt than for textile carcase belts. The service factor is generally around seven, but varies according to the belt and application.

ISO 15236-1 divides belts into two types, 'standard type' which does not have transverse reinforcement and belts that do have this reinforcement, which are simply referred to as 'conveyor belts having transverse reinforcement'. This seems rather odd terminology: calling something standard implies that others are not standardised. It specifies that steel cord belts shall be produced in strengths of between 500 N/mm and 8000 N/mm and provides a table of preferred belt strengths. Belt strength is defined as:

$$\frac{\text{Cord strength} \times \text{number of cords}}{\text{Belt width}}$$

The table gives three ranges of belts: 'Low', (500 to 1600 N/mm), 'Medium', (2000 to 3150 N/mm) and 'High' (3500 to 5400 N/mm). Belts with substantially higher strength ratings than 5400 N/mm have been produced. As described elsewhere in this book, a steel cord belt with a tensile strength rating of 7100 N/mm was used on a twelve kilometre long, single run installation at Gascoigne Wood mine in the United Kingdom. The standard goes on to define the mechanical requirements for steel cord belts, which will be discussed in detail in Chapter 7.

In its definition of preferred belt types, ISO 15236-2 begins with the assumption that belts of the same nominal strength rating have cords of the same diameter and pitch, or at least the same carcase thickness. ISO 15236-2 does not categorise belts according to the two types given in ISO 15236-1. Instead it divides belts into three types A, B and C, according to the kind of vulcanised joint that may be applied, following the three kinds of joint specified in ISO 15236-4.[5.12] These are further divided into A1 and A2, B1 and B2, and C1 and C2. Types A1 and A2 extend to the highest strength ranges and are designed to be jointed by the highest strength type of joints. The ratio of cord spacing to cord diameter is lower for Type A2 than for A1, with thinner cords for a given belt strength. The Standard indicates that in contrast to Types A, belts of Types B1 and B2 have transverse reinforcement. The distinction between B1 and B2 lies in the elongation of the cords, with B1 having relatively high elongation. The strength of these types of belt extends to lower values than Type A1. Belts of Types C1 and C2 have tension members that are fabric-like in construction with warp consisting of brass or zinc coated steel cords. The longitudinal cords of Type C1 have higher elastic elongation than those of Type C2.

It has to be said that ISO 15236-1 and ISO 15236-2 are not for the casual reader, the uninitiated or the faint-hearted. The fact that there are two 'types' of belt defined, one in Part 1 and another in Part 2 is confusing. In addition, ISO 15236-2 states that 'Compared with belt Types A, belt Types B are built with transverse reinforcements as indicated in prEN ISO 15236-1:1996'. The reader could be forgiven, therefore, for believing that belt Types A1 and A2 specified in ISO 15236-2 might correspond to the 'standard type' in ISO 15236-1, having no transverse reinforcement. However, an example of belt designations (see later) in ISO 15236-1 clearly indicates that Type A2 can have transverse reinforcement!

As with fabric carcase belts, the properties of the rubber compounds used to make the belts may be varied to suit the application. ISO 15236-1 specifies four grades of cover, H, D, L and K, based on their tensile strength, elongation at break and abrasion resistance. ISO 15236-2 specifies minimum cover thicknesses for the preferred belt types, which vary from three millimetres for the lightest belts to eight millimetres for the strongest and heaviest. Fire resistant grades of steel cord belt are available to meet the most stringent requirements.

ISO 15236-1 specifies a marking and designation system for steel cord belts. As with fabric belts, steel cord belts are usually marked by 'branding', which embosses the belt designation into the carrying cover of the belt. The designation is required to include the belt width, an indication that it is a steel cord belt, its strength designation, the existence of any weft reinforcement, the cover thicknesses and the belt type according to ISO 15236-2. Thus the designation

1 200 ST 5 000/10 + 7 H + A1

indicates that the belt was 1200 mm wide, was a steel cord belt (ST) of tensile strength 5000 N/mm with a ten millimetre cover on the carrying side and a seven millimetre cover on the pulley side, that the covers were of grade H and that the belt was of Type A1 according to ISO 15236-2. For belts with transverse reinforcement the letter 'T' indicates textile reinforcement and is included with the cover thickness values, whereas the designation 'S' indicates steel weft reinforcement and is included after the 'ST' steel cord designation. The distinction is made because textile weft reinforcements are considered to be part of the covers, while steel wefts are part of the core. Thus

1 000 ST 500/6T + 3T L + B1

indicates that the belt has textile breakers above and below the cords, while

1 000 ST S/S 500/6 + 3 L + B1

would indicate that steel reinforcement of the weft was in place both above and below the cords of the core. In addition to this information, the marking required on the belt must also include reference to ISO 15236-1 and the last two digits of the year of its issue, the belt manufacturer's name, the last two digits of the year of manufacture and the makers identification number for the belt.

5.4 Cable belts

As was explained in the introduction to this chapter, the essential feature that separates the cable belt design from the other types of bulk materials carrying conveyors is that the carrying element and the tension transmitting elements are separated. The belt, which is under very low tension, rides on endless steel wire ropes that transmit the motive power from the drive unit. It has no longitudinal reinforcement, because it is not the tension element, but does have lateral reinforcement in the form of straps or wires in the rubber belt core. This reinforcement is needed to assist in supporting the load because the belt is not supported by idlers across its width in the conventional way, but has vee-shaped 'shoes' towards the outer edges of the belt that ride on the wire ropes. The transverse support may be preformed to produce a convenient troughed shape to increase the volume of bulk material that can be carried, but in the past, cable belt installations seen by the authors have been flat. Top and bottom covers are applied to the reinforced core and the pairs of 'shoes' that run continuously along the entire length of the belt and ride on the wire ropes, are added to both top and bottom covers (Fig. 5.9). The whole assembly is vulcanised together. The pairs of wire ropes that provide the motive power for the conveyor belt are each supported on pairs of pulleys sited at intervals of six metres on the top strand and twelve metres on the bottom. Because the belt is not transmitting power, it avoids most of the design constraints that apply to textile and steel cord carcase belts. The choice of rubber covers and their thickness can be made to suit the environment in which the belt is to work, for example the belt can be made fire resistant if needed. In addition the length of a single flight is not constrained by the strength properties of the belt, and very long single flight conveyors can be installed. For example, Cable Belt Ltd claimed the world's longest conveyor in 1967 with an installation five and a half miles long at Longannet mine in Scotland. The installation conveyed the output of the mine to the Longannet power station. Because of the transverse reinforcement, this type

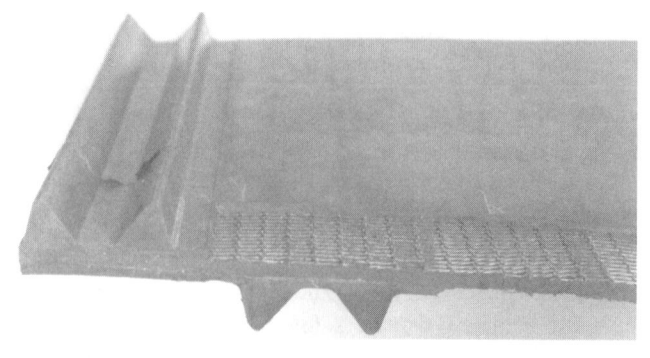

5.9 Cable belt construction (former NCB photograph – published by kind permission of Department for Business, Enterprise and Regulatory Reform).

of belt is less prone to rips than is steel cord belt. The separation of the carrying and drive elements gives greater flexibility in the negotiation of vertical and horizontal curves than other types of belt allow, since no account has to be taken of belt stresses and the appropriate guidance of the drive cables is more easily accomplished.

5.5 References

5.1 ISO 14890:2003 'Conveyor belts – Specification for rubber and plastics covered conveyor belts of textile construction for general use'.
5.2 'Accident at Cresswell Colliery, Derbyshire' Report by Sir Andrew Bryan Published by The Ministry of Fuel and Power, June, 1952 HMSO London.
5.3 BS 8438:2004 'Troughed belt conveyors – Specification'.
5.4 ISO 10247:1990 'Conveyor belts – Characteristics of covers – Classification'.
5.5 ISO 37:2005 'Rubber, vulcanised or thermoplastic – Determination of tensile stress-strain properties'.
5.6 ISO 4649:2002 'Rubber, vulcanised or thermoplastic – Determination of abrasion resistance using a rotating cylindrical drum device'.
5.7 BS 6593:1985 'Code of practice for on-site non-mechanical jointing of plied textile and steel cord reinforced conveyor belting'.
5.8 Gilbert S, Discussion to 'A laboratory study of the dual-drum multimotor conveyor drive' by Firbank T C *The Mining Engineer* No 140 May 1972.
5.9 EN 12882:2002 'Conveyor belting for general purpose use – Electrical and flammability safety requirements'.
5.10 ISO 15236-1:2005 'Steel cords conveyor belts – Part 1: Design, dimension and mechanical requirements for conveyor bets for general use'.
5.11 ISO 15236-2:2004 'Steel cords conveyor belts – Part 2: Preferred belt types'.
5.12 ISO 15236-4:20047 'Steel cords conveyor belts – Part 4: Vulcanised belt joints'.

6
Joining conveyor belts

6.1 Introduction

The joining of textile and steel cord carcase conveyor belts is a key area of conveyor technology. No matter what properties are engineered into the belt, the joints can present a serious weakness to the performance of the system. Like the belt, they have to withstand the tensions imposed during the passage round the conveyor and the differential tensions at transitions or vertical or horizontal curves and must possess the flexibility to pass around all of the system. In addition they must not damage or be damaged by belt cleaning systems, idlers, pulleys or any other part of the conveyor system. They must also be properly square to the axis of travel of the belt, otherwise the tracking of the belt may be affected.

Two systems of belt joining have been developed. These have been termed mechanical and chemical systems, although this is an oversimplification. Mechanical fasteners join the belt ends together by some form of hardware, whereas in chemical systems the belt ends are in essence bonded together. These 'chemical' joints are referred to generally as 'spliced joints', although confusingly, joints made by mechanical fasteners are sometimes known as 'mechanical splices'. Here we shall use the terms 'mechanical joints' for those made with hardware and 'splices' for those made by bonding. We shall deal with the two systems in turn.

6.2 Mechanical fasteners

Mechanical fasteners may be used to join both plied and solid woven types of fabric carcase conveyor belts. With this form of joint, hooks, staples, bolts or rivets are driven through the belt and carry attachments that may be used to secure the two belt ends together. These attachments may be interlocking loops or hooks that form a hinge by the insertion of a pin, or take the form of solid plates which are secured directly into both belt ends. In the very early days of conveyor belting, lengths of belt were joined by crude clips that consisted of

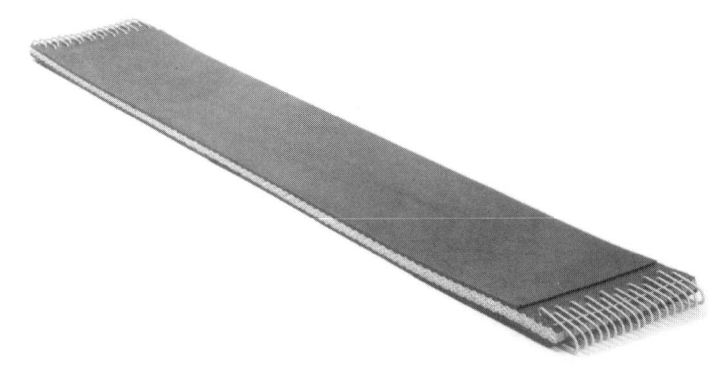

6.1 Wire hook fastener (former NCB photograph – published by kind permission of Department for Business, Enterprise and Regulatory Reform).

small plates with sharp prongs that were placed over the two belt ends. The prongs were driven through the belt using a hammer and the prongs were then clenched over on the underside. In a development of this, two sets of pronged plates were driven in, one on each side of the joint and then interlocked and joined with a pin to form a hinge. This joint had the advantage that it could be disconnected. Later it was common to use plate fasteners secured to the belt by rivets or bolts. The joints were time-consuming to make and damaged the belt considerably.

During the 1930s the wire hook type of fastener, developed in Germany and inserted by a machine, came into widespread use. Figure 6.1 shows a series of these hooks fitted into a piece of belting to form one half of a hinge joint. When synthetic fibres were introduced into belts their strengths increased substantially, but full advantage could not be taken of these strength increases because the strength of the joints available had not increased in proportion. Barclay[6.1] describes the development and introduction of the staple fastener, which was developed to rectify this problem. The staple fastener consists of a series of strong loops that fit over the belt. Staples pass through holes in the loops and then through the belt and are clenched into recesses in the loops on the other side. A special 'lacing' machine is used for this process. The loops attached to the two belt ends are interlocked and joined with a pin to form a hinge joint in the usual way. Barclay states that there were five types of problem with wire hook fasteners:

1. If the wires were of too small a gauge, the hooks tended to open out under the higher tensions imposed on the stronger belts.
2. If the wires were too large they damaged the belt, causing a line of general weakness and tearing from the back of the joint.
3. The hooks could pull through the belt if the belt had poor holding properties.
4. Some instances of hooks fracturing had occurred and were thought to be due to poor steel quality or heat treatment, or to fitting heavy duty fasteners on

belts that went round small diameter pulleys, thus inducing high stresses in them.
5. The wires could wear through, which should be the failure mode after long service.

The staple fastener design sought to eliminate some of these problems by optimising the staple wire diameter so that the staples did not open out or cause tearing of the belt and making the loops sufficiently robust to avoid them opening out under high tension. The staple fasteners improved the joint 'efficiency', i.e., the ratio of the joint strength to the belt nominal strength in a straight tensile pull, by 15 to 20%, from around 65 to about 80%. In field trials, the life of joints with staple fasteners was on average three times that with wire hooks. It should be noted here that the belts that Barclay used for his tests were of much lower strength than those in common use in the mining industry today, with tensile strengths in the range 2500 to 3500 lb/in (438 to 612 N/mm), hence higher figures for joint efficiency could be obtained.

Figure 6.2 is an example of a hinged joint and Fig. 6.3 an example of a plate joint. Plate joints are generally stronger than hinged joints, although the latter can be used with smaller pulleys. Suppliers of mechanical fasteners produce a considerable range of individual fastener sizes and types to cater for the different thicknesses and strengths of conveyor belt produced. In addition they supply machines to be used for the attachment of the various fastener types. Plate fasteners are commonly used in the USA with rubber covered plied belting, but are much less popular in other parts of the world. It is believed that the rubber covered ply belts act to cushion the plate fasteners as they pass round pulleys in a way that,

6.2 Hinged joint mechanical fastener (by kind permission of Flexco).

6.3 Plate joint mechanical fastener (by kind permission of Flexco).

say, PVC covered solid woven belting would not. It is, of course, essential that the right type and dimension of fastener be used for the belt. If the staples, rivets, or bolts are either too long or too short then an inferior joint will be produced. Equally, the right machine is needed for the insertion of the staples or rivets.

Preparation of the belt ends to ensure that they are perpendicular to the centre line of the belt is essential to avoid uneven distribution of tension across the joint. It is common practice to chamfer the edges of the belt to make the joint slightly narrower than the belt itself to avoid damage to the edges of the joint and belt. The slight disadvantage of this practice is that it increases the stress across the joint. Where cover thickness permits it is useful to remove part of the cover thickness in the area of the joint so that the main part of the fastener sits slightly below the belt surface, minimising the risk of damage to the joint or to other parts of the conveyor (see Fig. 6.1).

The principal advantage of mechanical joints is that they can be made quickly, by relatively unskilled personnel, with relatively little in the way of specialised equipment. A degree of spillage of the material being transported is inevitable with mechanical fasteners, but the principal disadvantage is that the strength of the joint is well below the strength of the belt itself. From the discussion in Chapter 5 about belt constructions, it will be clear that although the belt strength is determined by the strength and number of the longitudinal warp threads, it is the transverse weft threads that are important in determining the joint strength by their ability to retain the staples, rivets or bolts that pass through the belt. Therefore, regardless of the strength of the metallic components of the fasteners, the construction of the belt is a vital factor in determining the strength of a joint. Barclay[6.1] found that for a given overall tensile strength of belt, fastener holding was better for a close weave than for an open one. He also found that for equal strengths in the weft (transverse) direction fastener holding was better if the weft had a large number of weaker threads than if it had a smaller number of stronger ones.

The former British Coal Corporation had a requirement that new mechanical joints in new belting should have a strength in a tensile pull of 50% of the strength of the belt. After fatigue cycling to a specified regime, simulating service, the strength of the joint was required to be 40% of belt strength. (More details of these tests and requirements are given in Chapter 7.) Textile belting is almost universally rated at a service factor or factor of safety of 10:1, i.e., the maximum running tension is one-tenth of the tensile strength of the belt. If the strength of a joint that has been on the conveyor for some time is 40% of belt strength then it will withstand four times the maximum running tension. However, if we bear in mind that a motor driving through a fluid coupling can supply up to three times full load torque to the belt if the belt is stalled, then we can see that with the effects of wear in service, or poorly made joints, or both, there is not a great margin of safety.

The problems related by Barclay concerning the need for improved joint strength following increases in belt strength were revisited more recently by Clamp[6.2] who has reported on experience relating to the performance of belt joints on conveyors used in the UK coal mining industry for man-riding, i.e., belts on which men are transported into and out of the mine. He relates that in 1995 a series of dangerous occurrences was recorded when mechanical joints on different man-riding conveyors failed independently. He ascribes the problem to the fact that while the strength and quality of conveyor belting had evolved, the strength of the fasteners had not improved in proportion. He gives figures. In former years belts commonly used in mines were solid woven Type 6, rated at a tensile strength of 6000 lb/in (equivalent to 1050 N/mm) and new mechanical joints had a tensile strength of 65% of belt strength, 3900 lb/in (683 N/mm). Belts had been upgraded to solid woven Type 8, 8000 lb/in, (1400 N/mm), but joint strength was now typically 55% of belt strength, 4400 lb/in (770 N/mm). Thus while the belt strength had increased by a third, the joint strength had increased by only 13%. Investigations showed that the main cause of joint failure was not the fastener itself but by the joint 'combing out', i.e., the weft threads failing to hold the fasteners. A development programme was undertaken to improve the fastener holding properties of the belting to the Type 6 level, i.e., 65% of belt strength. Clamp goes on to quote the results of examining joints on the conveyors that had suffered failed joints. He cites the following 'fundamental and elementary' problems:

- ends of belting not cut square
- cutting into, and hence weakening, the fabric during preparation of the belt ends

- use of the wrong type of fasteners for the belt thickness
- fasteners not pushed fully back before being riveted or stapled
- fasteners had not been fully riveted or stapled to grip the belt properly.

He reports that, together with an education programme to improve the quality of joints made, the joint strength development work was successful in reducing drastically the number of joint failures experienced.

All mechanical joints that will fit into a given belt will not give equivalent performance; neither will they all be suitable for the same pulley diameters. Advice should be sought from both belt and fastener suppliers on these points.

6.3 Spliced joints

Spliced joints are generally installed by specialist splicing organisations, requiring rather more skill and more equipment than mechanical joints, but they provide much stronger and more permanent joins in belts. They offer reduced maintenance, reduced spillage, reduced noise and reduced idler wear compared to mechanical joints. On the downside, they are more expensive to install and require longer downtime for their installation. Different techniques are applicable to textile carcase and steel cord belts. Three standards, BS 6593 'Code of practice for on-site non-mechanical jointing of plied textile and steel cord reinforced conveyor belting',[6.3] DIN 22102-3 'Conveyor belts with textile plies for bulk solids – vulcanised joints'[6.4] and ISO 15236-4 'Steel cords conveyor belts – Part 4: Vulcanised belt joints',[6.5] all offer guidance on the design and execution of splices.

6.3.1 Textile carcase belts

Plied textile carcase belts may be joined by a variety of types of splice, variously known as the step splice, the scab splice, the butt splice, the skive splice, the jump type splice and the finger splice. Not all of these are suitable for all belt types. They may be divided broadly into two generic classes: joints in which the ends of the plies are either butted together or overlapped in some way, and ones in which the belt ends are cut into triangular 'fingers' that are interleaved.

The step splice is shown in Fig. 6.4. During the making of this type of splice the plies are cut away to form steps that mate up with matching steps made in the other belt end. The steps are usually cut at an angle to the longitudinal axis of the belt. This angle, which is known as the bias, is 20–22 degrees. This type of joint is sometimes made with the steps cut in a chevron pattern, which is said to minimise edge damage. The step lengths are important in giving sufficient bonding area to enable the joint to develop the required strength. BS 6593 gives details. The rubber interlayer between the plies is removed leaving a thin layer to assist bonding. The interlayer is replaced by the bonding medium, which is

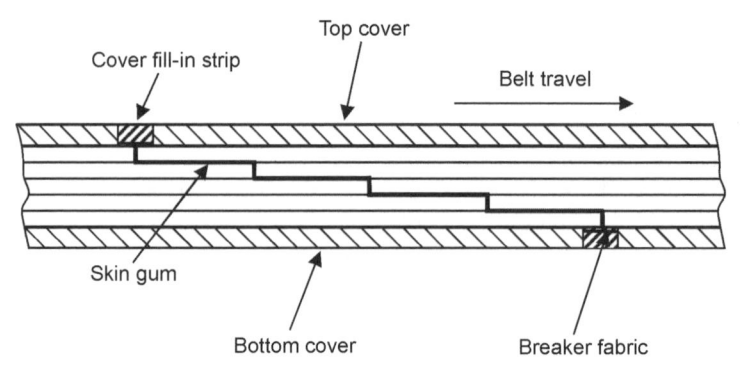

6.4 Step splice in plied belting.

referred to as the skim gum (or skin gum in BS 6593). The trailing edge of the top ply faces away from the direction of travel of the belt to minimise damage to the joint by belt scrapers or ploughs. For general purpose belts not having special covers, BS 6593 recommends vulcanising at 150 °C at a minimum pressure between the press platens and the belt surface of 700 kPa. The vulcanising period depends on the belt thickness and varies from 25 minutes for belts up to and including 6 mm thick, to 45 minutes for belts between 20 and 26 mm thick.

The splicing procedure for PVC impregnated plied belts is much the same as is given above except that the skim gum is replaced by PVC paste and the joint is bonded together by a gelling process rather than by vulcanising.

The scab splice or butt splice is used on mono- and duoply belts and could be considered to be a version of the step splice for these two types of belt. Where there is a single ply, the two ends to be joined are butted together and a high strength fabric – the scab fabric – is bonded across the joint. It is this fabric that is the tension transfer medium. In the case of a two-ply belt there is a single step, and the scab fabric is bonded over either the top or the bottom of the joint or both. The skive splice is a name given to a butt splice where the belts' ends are skived or cut at an angle of 45 degrees when the belt is viewed edge on. The jump splice is another variant of the scab splice but in this case parts of the plies are overlapped, so that the joint is thicker than the belt.

Where heating is not available the cold cure method is sometimes used, where the bonding agents will cure at ambient temperatures. However, pressure is still required to achieve satisfactory results.

The principal area of application of the finger splice (Fig. 6.5) is in the jointing of PVC and rubber impregnated solid woven belts, although finger splices may be used on plied belts and on some types of steel cord belt also. Belt manufacturers have developed splicing techniques over many years and have derived rules for the aspect ratios (lengths and widths) of the fingers to give high joint strengths with adequate flexibility to avoid possible tracking problems. In

6.5 Finger splice (by kind permission of Fenner Dunlop Conveyor Belting Europe).

the authors' experience joint strengths in solid woven PVC impregnated belting are regularly over 75% of belt strength and can be as high as 100%. Both polyurethane and PVC are used as the bonding medium. In the UK during the 1980s and 1990s belt manufacturers supplying to the nationalised coal mining industry switched away from polyurethane that had traditionally been used, to the use of PVC as the bonding medium. This switch followed pressure from the coal mining industry to eliminate the use of polyurethane. This pressure arose because of health fears relating to sensitisation of personnel by isocyanates present in one of the constituent chemicals of the polyurethane. This sensitisation can lead to asthma and similar breathing problems. Polyurethane is now in use again as the bonding medium for joints in the now privatised UK coal industry in some circumstances and has continued to be used in some power generation plants and in potash mines in the UK and overseas.

The sequence of events involved in the making of a finger splice in solid woven PVC belt, using PVC as the bonding medium, starts with the drawing of the fingers and their cutting, following which the covers are removed from the fingers. The bottom rubber cover is then positioned in the press and smeared with PVC paste. A breaker fabric is placed over this and more paste applied. The fingers are then positioned over the breaker and bedded in. More PVC paste is applied followed by the top breaker and the top cover. The press assembly is completed and the whole joint is then cured under heat and pressure. Joints in rubber impregnated solid woven belting are made in much the same way, except that rubber is used as the bonding medium and small gaps are left between the fingers when the joint is laid up to allow the rubber to penetrate. The use of the breaker fabric is important in protecting the joint and equalising the distribution of stresses across the joint. Accurate control of the temperature and pressure at which the joint is cured are vital for the making of sound joints.

6.3.2 Steel cord belts

In our discussion of steel cord belts in Chapter 5 we referred to preferred belt types, as given in ISO 15236-2,[6.6] which defines belt types by the kind of vulcanised joint that may be applied. These vulcanised joints are specified in ISO 15236-4,[6.5] which divides them initially into two types, stepped joints and finger joints. Stepped joints are further divided into interlaced stepped joints or plain stepped joints. Since we have discussed finger joints in the previous section and since the principles and techniques used for finger joints in textile belts apply in general to those types of steel cord belt for which finger joints may be applicable, we shall concentrate here on stepped joints.

The difference between interlaced stepped joints and plain stepped joints is that the former contain more cords than the original belt, while the latter contain the same number of cords as the original belt. Figures 6.6 and 6.7 illustrate these two types of joint. It will be clear from these figures that the cords in the belt ends are cut according to particular sequences and that the cuts are staggered to produce 'steps'. Splices may have 2, 3, 4 or more steps, the pattern of which is repeated across the belt. The precise design of a joint in terms of the sequence of cord cuts, the length of the splice and the number of steps is left to the belt manufacturer or to the company requiring the splice. However, ISO 15236-4 gives some useful basic information regarding splice design and execution. The quality of the bonding between the cords and the rubber is, of course, vital if the joint is to be able to transmit the forces imposed on the belt from one belt length to the next. A key feature of joint design is that there should be sufficient rubber between the cords to allow adequate bonding. BS 6593 recommends 1.5 mm as a minimum, whereas ISO 15236-4 relates the minimum gap between cords to the cord diameter, using the formula

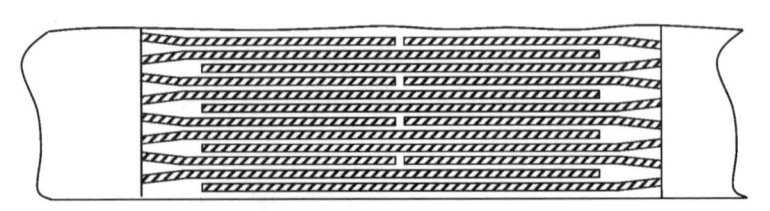

6.6 Lay-up of an interlaced spliced joint.

6.7 Plain stepped joint: top – organ pipe; bottom – fir tree.

Gap $\geq 1.2 + (0.1 \times \text{cord diameter})$ millimetres

It is important that when the cords are stripped from the original belt rubber some rubber remains attached to them to assist in the bonding to the filler rubber inserted into the joint. Some manufacturers provide preforms for this filler rubber into which the cut and stripped cords are laid during the laying up of the joint. The use of a breaker fabric in the joint is important in equalising the distribution of stresses across the belt joint. Temperature control is again vital to the production of satisfactory joints. A joint failed on a major installation with which the authors were concerned because it was not cured at the correct temperature. The vulcanising press had two thermocouples to indicate the temperature of the joint. One was defective, but unfortunately the jointing team chose to believe the wrong one!

6.4 Concluding remarks

As explained in the introduction to this chapter, belt joints are a key feature of conveyor technology and a potential Achilles heel. We have seen that the technology of mechanical fasteners is relatively unsophisticated, but that even so the fitting of these fasteners is not without pitfalls. We have also seen that improvements in the strength of belts must be accompanied by improvements in the performance of the fasteners for there to be any real gain. We have also seen related instances where fastener technology has had to 'play catch-up'. The joining of belts seems to be an area of conveyor technology that has been relatively neglected considering its importance. It deserves more attention.

6.5   References

6.1   Barclay J T 'Conveyor belt joints and fasteners' *The Mining Engineer* 126 June 1967 p. 612.
6.2   Clamp G 'Transport safety in RJB' *Proceedings of Institution of Mining Engineers Symposium on Developments in Underground Transport* 1998.
6.3   BS 6593:1985 'Code of practice for on-site non-mechanical jointing of plied textile and steel cord reinforced conveyor belting'.
6.4   DIN 22102-3 'Conveyor belts with textile plies for bulk solids – vulcanised joints'.
6.5   ISO 15236-4:2004 'Steel cords conveyor belts – Part 4: Vulcanised belt joints'.
6.6   ISO 15236-2:2004 'Steel cords conveyor belts – Part 2: Preferred belt types'.

7
Standards, test methods and their standardisation

7.1 Introduction

So far in this book we have discussed conveyor design, belt construction and jointing. We have appreciated the need for tension control and the subtleties of belt manufacture, and have seen that belt joints can be an area of significant problems. It is clear that, although a conveyor is nominally a fairly straightforward device, there are many things that, potentially, can go wrong. As with any piece of machinery, for a conveyor to work properly we need to be assured not only that the design is right, but also that all of the elements perform in the way that they are expected to do. We want to know that the elongation of the belt is going to be what we designed for when we put in the take up, that the plies are not going to come apart after a couple of passages round the conveyor, or that the strength of the belt joints is adequate. Much of our assurance that all will be as it should be comes from type and quality control testing that has been carried out to determine and confirm the properties that we need for the design and construction of the conveyor. The purpose of this chapter is twofold: to discuss the type of tests that are applied to conveyor belting to ensure that the expected properties and performance are actually achieved, and to provide information on the standards that have been laid down for these tests. We shall concentrate here on the conveyor belting. Most of the rest of the components of a conveyor are used in other types of equipment and there seems little point in discussing the testing of motors, ball bearings or fluid couplings, for example, when the novel, and key item in the conveyor is the belt itself.

This chapter does not consider tests to examine fire safety performance or the propensity of belts to produce static electricity. These matters are considered to be of sufficient importance to be dealt with in depth in Chapter 8.

7.2 General remarks

Many of the tests that are currently applied to conveyor belting originated in quality control tests. Others arose because designers needed to have certain data or because purchasers needed to be assured that they were getting what they had ordered. This chapter is not intended to cover every conceivable test that was ever devised for conveyor belts, but to give an overview of the principal areas of importance. The standardisation of tests within an industry area is a natural development that is beneficial to manufacturers, designers and purchasers. Many countries have produced national standards for conveyors and conveyor belting, but two particular bodies need to be mentioned and the relationship between them, and between them and the national bodies, explained. These two bodies are the International Organisation for Standardisation (ISO) and the European Committee for Standardisation (CEN). We shall concentrate wherever possible on tests that have been standardised by ISO because the majority of the countries of the world has accepted them and by the CEN since this embraces the twenty-seven countries of the European Union. In some instances where there is no ISO or CEN standard, British or German standards are mentioned for guidance. Appendix 2 provides a source of reference for standards relating to conveyors, containing a listing of the majority of ISO and CEN standards that are relevant to conveyors for bulk minerals transport. It also lists many national standards.

7.3 The standardisation process

Because many of the tests currently made on belts originated as quality control tests developed by belt manufacturers, little of the development work is documented. In some instances, tests were developed by particular industries to suit their special requirements, particularly where, as in the UK, some of the large industries such as coal, steel and power generation were nationalised. Standardisation is required to allow proper comparisons to be made between products from different suppliers, and to ensure quality, reliability, interchangeability and safety. Users, manufacturers, trade associations and similar bodies come together under the auspices of the national standardisation body to produce these standards. One of the earliest conveyor belt standards, British Standard 490 'Rubber conveyor and elevator belting' was first published in 1933 and contained tests for various physical properties such as tensile strength and adhesion between plies and of covers to plies. This standard was originally written on the assumption that the fabric carcases of all belts were made exclusively from cotton. With the introduction of synthetic fibres into belt construction and the use of PVC as well as or instead of rubber, the tests and requirements in BS 490 had to be revised.

The requirements introduced by the coal mining industry in Britain for belts used underground to be fire resistant and anti-static resulted in a new standard BS 3289 'Conveyor belting for use in underground coal mines' being introduced in 1960, following the production by the National Coal Board of its own standard number 158 'Fire Resistant Conveyor Belting'. Both BS 3289 and NCB

158 drew upon the existing BS 490 but revised and added to the tests and requirements in that standard. The British Steel Corporation, now part of Corus, produced its own standard CES 8 for conveyor belting used in the UK steel industry, while in power generation the tendency has been to use the Canadian standard CAN/CAS-M422-M87, which is basically a safety standard. Some test development work is reported by Barclay[7.1] but that mainly concerns tests that fell outside the scope of BS 490 and BS 3289, i.e., flexibility, coefficient of friction, ageing characteristics and water absorption. Manufacturers are still, of course, developing and using for quality control and to ensure fitness for purpose, additional tests that have been developed 'in house' and have not been standardised either nationally or internationally (see for example ref. 7.2, the *Fenner Dunlop Technical Manual*).

ISO is the world's largest developer of standards. It is a non-governmental organisation that is composed of the national standards institutes of 157 countries. International standardisation began with the formation of the International Electrotechnical Commission (IEC) in 1906. In 1926 the International Federation of the National Standardising Associations (ISA) was set up, but its activities came to an end in 1942. In 1946, delegates from 25 countries met to form a new international organisation, 'to facilitate the international coordination and unification of industrial standards'. The new organisation, ISO, began operations on 23 February 1947. The British Standards Institution (BSI) took the approach of adopting ISO standards where this was possible and putting the national letters BS in front of the ISO in the designation of the standard, so that ISO xxx became BS ISO xxx.

CEN was founded in 1961 by the national standards bodies in the European Economic Community (EEC) and European Free Trade Area (EFTA) countries. With the transformation of the EEC, by a process of mission creep, into the European Union (EU), came the agreement by member countries that EU Directives would be enacted into their national laws. Since the European Union seeks to promote free trade, enhance the safety of workers and consumers and protect the environment, Directives have been produced to facilitate these ends. National standards were seen as presenting barriers to free trade, and standards that could be agreed across Europe were needed. Some of the Directives concern what are termed Essential Health and Safety Requirements (EHSR), compliance with which is compulsory for all manufacturers, suppliers and users of machinery. One method of showing compliance with the EHSR is to manufacture and use equipment in accordance with 'harmonised' European standards specifically drawn up to help manufacturers meet EHSR. The role of CEN is thus principally one of producing harmonised standards that support EU Directives, i.e., that provide a means by which compliance with the Directives may be achieved. Complying with harmonised CEN standards should mean that a product will meet all EU Directives applicable to it. CEN has produced a hierarchy of harmonised standards:

- Type A standards set out the basic concepts that can be applied to all machines.
- Type B standards each deal with one particular safety aspect or device that can be applied across a wide range of machines.
- Type C standards give detailed safety requirements for a particular machine or group of machines.

All of the national standards bodies in the EU are obliged to withdraw any conflicting national standards once a CEN standard has been produced. The national standards-making bodies of Europe have adopted the device of putting their national designation letters in front of the European designation EN. The European standard EN xxxx may therefore be referred to as BS EN xxxx in Britain or DIN EN xxxx in Germany.

In some areas of their activities ISO and CEN co-operate closely, so much so that an agreement was reached between them that where one body was working on producing a standard on a particular topic, the other would not need to work on that area. There was a great deal of sense in this arrangement, because as far as Europe was concerned, the same people were sitting on both CEN and ISO committees. Some standards that are accepted as being technically equivalent under this agreement are dual numbered. Thus the standard EN xxxx may also be ISO xxxx and numbered BS EN ISO xxxx in Britain, or DIN EN ISO xxxx in Germany.

A feature that many people find annoying is that ISO and CEN put test methods in one standard and the requirements, or values that should be met, in another. So for any test, two standards are needed, one to tell you how to do the test and another to tell you what values are satisfactory. This does not endear the standards-making bodies to the user, but does sell more standards! It is understood that the original intention was for the test methods to form Part 1 of each standard and for the requirements to form Part 2. Many of the test methods were produced as Part 1s. However, the requirements (Part 2s) have been brought together into the product standards. The intention now is to revise the numbering of the test methods so that the 'Part 1' designation disappears and, for example, ISO 283-1 will become simply ISO 283. In the comments given in this chapter on the various standards reviewed, requirements are given where it is appropriate to do so.

7.4 Specific standards and tests standardised by the International Organisation for Standardisation (ISO) and the European Committee for Standardisation (CEN)

The following four product standards contain the general constructional, physical and mechanical requirements for conveyor belts:

1. ISO 14890:2003 'Conveyor belts – Specification for rubber or plastics covered conveyor belts of textile construction for general use' (EN ISO 14890:2003).
2. ISO 15236-1:2003 'Steel cord conveyor belts – Part 1: Design, dimensions and mechanical requirements for conveyor belts for general use' (EN ISO 15236-1:2003).
3. ISO/DIS 15236-3:2005 'Steel cord conveyor belts – Part 3 – 'Special safety requirements for belts for use in underground environments'. (DIS = Currently in course of preparation. Will become EN ISO 15236-3.)
4. ISO/DIS 22721:2005 'Conveyor belts – Specification for rubber or plastics covered conveyor belts of textile construction for underground mining'. (DIS = Currently in course of preparation. Will become EN ISO 22721.)

The following sections provide references to the standards that describe the main physical or mechanical tests on belts and comment on the salient points regarding these tests.

7.4.1 ISO 283-1:2000 'Textile conveyor belts – Full thickness tensile testing – Part 1: Determination of tensile strength, elongation at break and elongation at the reference load'

Dumb-bell-shaped test pieces, i.e., ones in which the centre section is narrower than the ends, are pulled to failure (Fig. 7.1). Test pieces are taken in both warp and weft directions and from the results the tensile strength in N/mm width are determined and the two elongations indicated in the title are calculated as

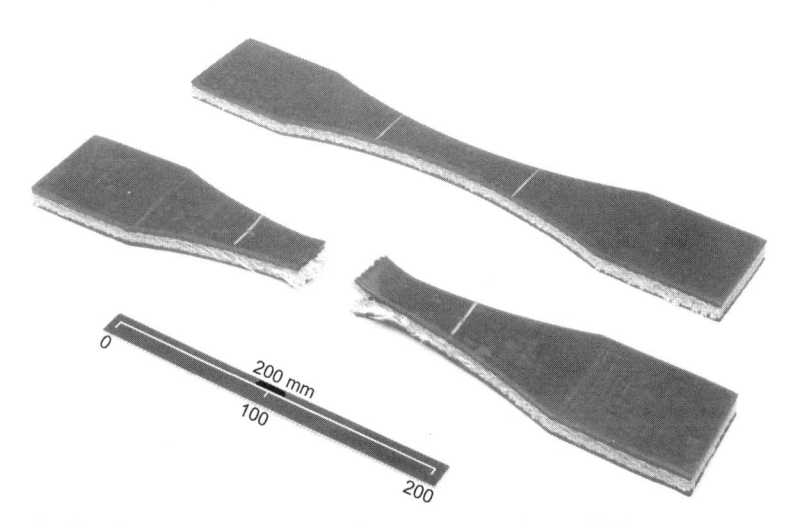

7.1 Tensile test pieces from a textile carcase belt (former NCB photograph – published by kind permission of Department for Business, Enterprise and Regulatory Reform).

percentages of the original test piece gauge length. The reference force is one-tenth of the nominal tensile strength of the belt, multiplied by the specimen width and is taken as such because the running tension for textile belts is most usually one-tenth of the tensile strength of the belt. The test pieces are waisted to try to ensure that breakage takes place within the narrowed gauge length. The test is applicable to belts with strengths up to 2600 N/mm: beyond this figure there is generally a difficulty in gripping the test piece adequately for the test to be accomplished. The requirements in ISO 14890 state only that the values obtained in this test should be not less than those set out for the particular designated belt types.

7.4.2 ISO 583-1:1999 'Conveyor belts with textile carcass – Total thickness and thickness of elements'

This test is used primarily for quality control purposes, but may be used by purchasers to check that they are getting what they paid for. It contains methods that are applicable to both plied and solid woven textile belts and describes how the thickness of the whole belt, the covers, the carcase and the plies may be measured. ISO 14890 does not specify either total belt thickness or cover thickness, but does give values for permissible tolerances on these measurements.

7.4.3 ISO 252-1:1999 'Textile conveyor belts – Adhesive strength between constitutive elements – Part 1: Method of test'

This standard describes the way in which the mean force required to strip the covers from the carcase and each ply from the next ply is measured. Test pieces of given dimensions are conditioned, i.e., held at given temperatures for a standard time. The cover or ply is separated from the rest of the belt at one end using a knife for a distance suitable for the ends to then be gripped in a motorised tensile testing machine. The test piece is pulled so that 100 mm of stripping is achieved (Fig. 7.2). Tests in both the warp and weft directions are made. The mean stripping force is calculated according to set procedures. (ISO 6133 'Rubber and plastics – Analysis of multi-peak traces obtained in determinations of tear strength and adhesion strength'). ISO 14890 gives minimum requirements for belts with and without natural fibres. The required values for adhesion both between plies and cover to carcass are some 30 to 50% higher for belts without natural fibres than for those containing natural fibres. The standard also gives limits on the variability of the values.

Standards, test methods and their standardisation 101

7.2 Ply adhesion test pieces (former NCB photograph – published by kind permission of Department for Business, Enterprise and Regulatory Reform).

7.4.4 ISO 9856:2003 'Conveyor belts – Determination of elastic and permanent elongation and calculation of elastic modulus'

As we have seen previously, the elongation is needed for the design of the take-up device and the elastic modulus for the calculation of transition distances. Test pieces are subjected to two hundred sinusoidal stress cycles between 2 and 10% of the nominal breaking strength of the belt sample. The graph of load against elongation shows both permanent and elastic elongation. The elastic modulus is calculated by dividing the applied force range (between 2 and 10%) by the relative elastic extension, i.e., the elastic extension divided by the original gauge length. In conveyor belting the elastic modulus is expressed in newtons per millimetre width, unlike in general engineering where it is expressed as force per unit cross-section. There are no requirements in ISO 14890.

7.4.5 ISO 505:2000 'Conveyor belts – method for the determination of the tear propagation resistance of textile conveyor belts'

An initial cut is made in a test piece and the resistance to tearing of a belt from this cut is measured (Fig. 7.3). The test may be useful for belts in applications where there is a risk of longitudinal tearing. ISO 14890 does not contain any requirements for this test, but for guidance it may be useful to note that the

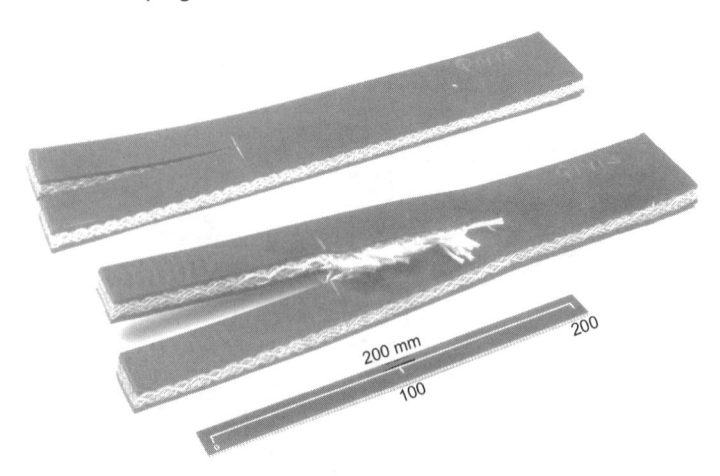

7.3 Tear propagation resistance of textile carcase belt (former NCB photograph – published by kind permission of Department for Business, Enterprise and Regulatory Reform).

former British Coal in its 'Specification for textile belts for use in coal mines' required tear strengths of between 1.0 kN for a belt of tensile strength 380 N/mm and 1.6 kN for a belt of strength 1400 N/mm.

7.4.6 ISO 1120:2002 'Conveyor belts – Determination of the strength of mechanical fastenings – Static test method'

The standard gives methods of testing for the types of mechanical fastener that are joined by a connecting rod (hinge type) and those that are not (plate type) (see Chapter 6). With the hinge type fasteners (Fig. 7.4) one-half of the joint is replaced by a steel connecting plate or 'comb', whereas with the plate type fasteners the complete joint is tested. ISO 1120 does not, of course, give requirements. However, both the forthcoming ISO and CEN standard EN ISO 22721 (see above) and BS 8407:2002 'Specification for mechanical and spliced joints in conveyor belting for use underground' do give requirements. Both require the minimum static fastener strength to be 60% of nominal belt strength for lower strength belts and a minimum of 50% of belt strength for higher strength belts. Where the standards differ is in the definition of lower and higher strength ranges. For EN ISO 22721 the lower range is up to a nominal tensile strength of 1250 N/mm, whereas for BS 8407 it is up to 1400 N/mm.

7.4.7 ISO 703-1:2000 'Conveyor belts – Transverse flexibility and troughability – Part 1: Test method'

A transverse section of belt is suspended and allowed to deflect under gravity. The troughability is expressed as the quotient of the deflection and the width of

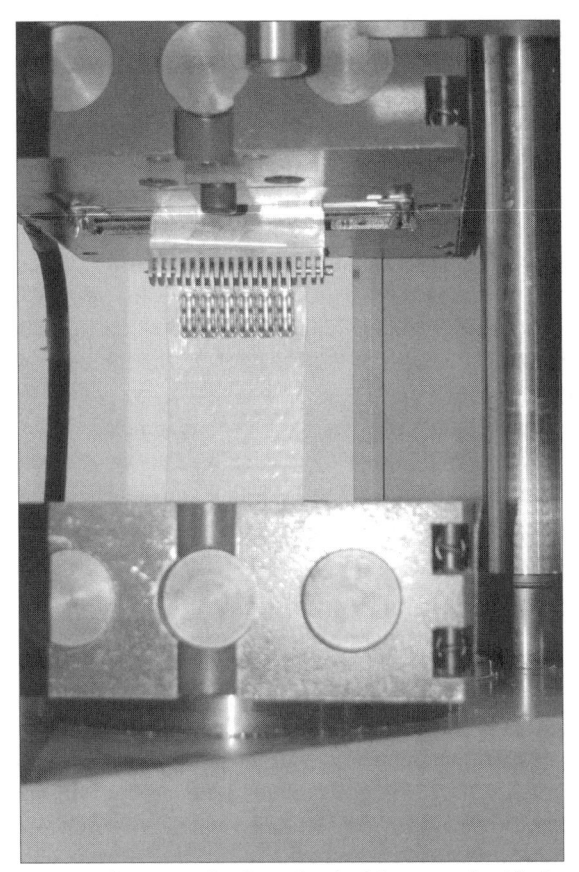

7.4 Tensile strength of mechanical fastener (by kind permission of Fenner Dunlop Conveyor Belting Europe).

the belt sample. The test is applicable to both textile and steel cord belts. Both ISO 14890 and ISO 15236-1 have the same requirements, which depend upon the troughing angle of the idler roller sets. For example, for angles of 25, 35 and 45 degrees the values are 0.10, 0.14 and 0.18, respectively.

7.4.8 ISO 7590:2001 'Steel cord conveyor belts – methods for the determination of total thickness and cover thickness'

This test is again used primarily for quality control purposes, but may be used by purchasers to check that they are getting what they paid for. Several methods are included, some of which allow for the presence of breakers or wefts in the belt construction. ISO 15236-1 states that for standard type belts (see Chapter 5) the covers should not be less than (0.7 × cord thickness) or not less than 4 mm, whichever is the greater. It gives no specific figures for belts with transverse reinforcement. This standard is currently being revised.

7.4.9 ISO 7622-1:1996 'Steel cord conveyor belts – Longitudinal traction test Part 1: Measurement of elongation' and ISO 7622-2:1996 'Steel cord conveyor belts – Longitudinal traction test Part 2: Measurement of tensile strength'

Both of these are essentially quality control tests. The elongation test is carried out on sample cords taken from a length of belt. The extension of the cords is measured at 10% and 60% of the specified minimum tensile strength. There are no specific requirements in ISO 15236-1. The tensile strength measured in Part 2 is the theoretical maximum strength of the belt, rather than the effective tensile strength, which will be lower, but which can be derived from the values measured. The test is carried out on a piece of belt containing five cords, four of which have been cut through to leave the centre cord to be subjected to the tensile test. The requirements in ISO 15236-1 are that the values obtained should be not less than the nominal belt strength.

7.4.10 ISO 7623:1996 'Steel cord conveyor belts – Cord-to-coating bond test – Initial test and after thermal treatment'

The force required to tear one of the cords out of the belt by applying a force along the axis of the cord is measured (Fig. 7.5). The thermal treatment referred to in the title of the standard is intended to simulate the effect on the bond of the heating during splicing. ISO 15236-1 gives requirements for the pull-out force of ($15 \times$ cord diameter $+15$) N/mm for the initial test and ($15 \times$ cord diameter $+5$) N/mm after thermal treatment.

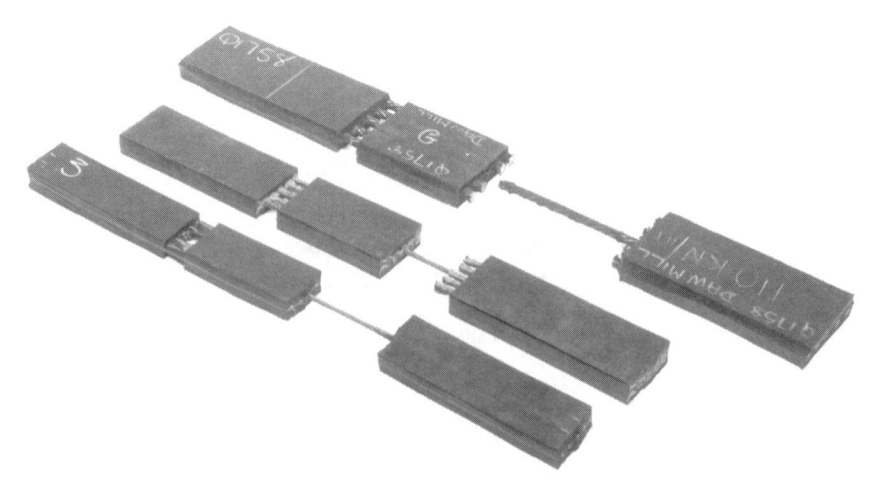

7.5 Cord pull-out test pieces (former NCB photograph – published by kind permission of Department for Business, Enterprise and Regulatory Reform).

7.4.11 ISO 8094:1984 'Steel cord conveyor belts – Adhesion test of the cover to the core layer'

This is basically a quality control test in which the covers are stripped from the core in a controlled manner. ISO 15236-1 requires a value of at least 12 N/mm of width.

7.4.12 Cover properties

The performance of the covers of belts is important to the user and both ISO 14890 and ISO 15236-1 give requirements for various grades of cover. The tests used to assess the cover properties are given in ISO 37 'Rubber, vulcanised or thermoplastic – Determination of tensile stress-strain properties' and ISO 4649 'Rubber, vulcanised or thermoplastic – Determination of abrasion resistance using a rotating cylindrical drum device'. However, both ISO 14890 and 15236-1 warn that reliable assessments of the performance in service of covers cannot be inferred from the results of the tensile and abrasion resistance tests alone.

7.5 Other tests

There are two other important areas of belt performance that are not covered by ISO or CEN standards. These concern the performance of joints in belts, and although ISO 1120:2002 describes the method for determining the static strength of mechanical fastenings, neither the static strength of spliced joints nor the dynamic strength of either mechanical or spliced joints have been addressed. There are, however, British and German standards that are relevant.

In British Standard 8407:2002 a method of testing is described that involves the pulling to failure of a loop of belting containing two vulcanised joints. For textile carcase belts vulcanised joints are required to have a minimum strength in this test of 60% of the full thickness tensile strength of the parent belt. The requirement for steel cord belts is 70% of belt strength.

With regard to the dynamic strength of joints in textile and steel cord belts, BS 8407:2002 gives methods of testing to assess the performance of both mechanical and vulcanised joints after a simulated service period. The resulting performance is referred to as the dynamic strength of the joints. Figure 7.6 shows a diagrammatic representation of the rig used for these tests. For mechanical joints, loops, containing four or six joints depending on belt strength, are run at a tension of 10% of nominal belt tensile strength on a rig containing four pulleys. The loop lengths and the pulley sizes are adjusted according to the belt type tested. Joints in belts up to 1000 N/mm tensile strength must survive 75 000 cycles, while those with nominal tensile strength greater than this must survive 100 000 cycles. Vulcanised joints must have strengths of at least 50% of the belt strength, but in this case after running on the test rig at 12.5% of belt strength.

106   Belt conveying of minerals

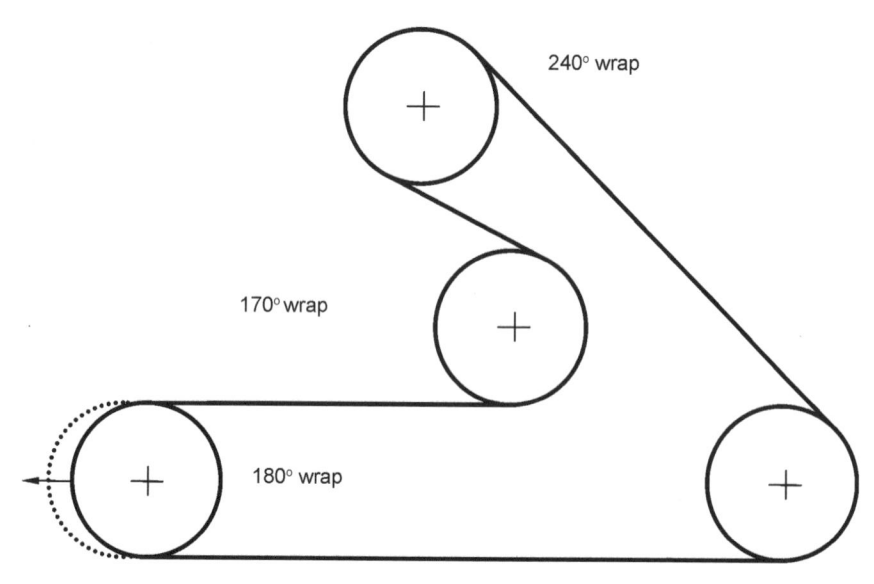

7.6 Four-pulley dynamic belt joint test rig.

For these joints the numbers of cycles are 100 000 for belts up to 1000 N/mm and 250 000 for belts above this.

The German standards DIN 22110-2 and 22110-3 (see Appendix 2) also define test methods for the dynamic testing of belt joints. The test rig described in DIN 22110-2 is shown diagrammatically in Fig. 7.7. As the figure shows, one

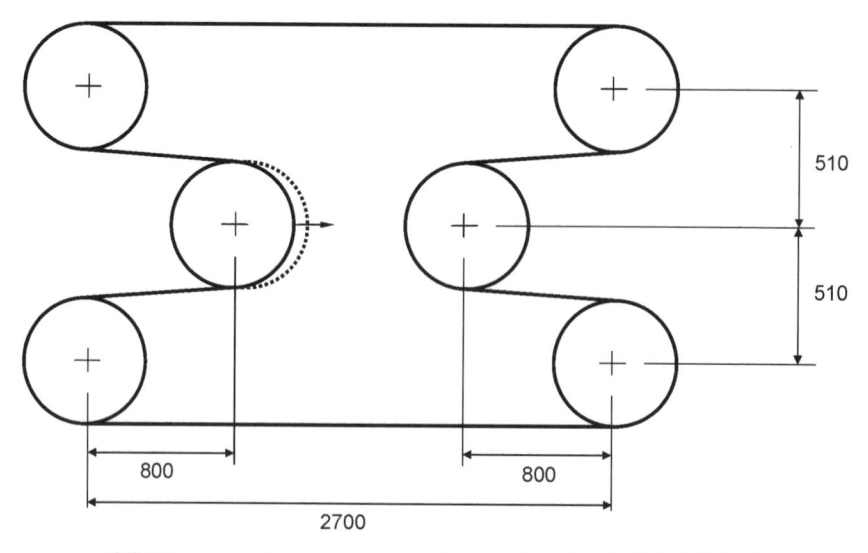

7.7 Diagrammatic representation of test rig described in DIN 22110-2 with 500 mm pulleys.

of the centre pulleys is moveable to provide tension to the belt. The diameter of the pulleys is 500 mm for belts with a tensile strength rating up to 1000 N/mm and 800 mm for belts with higher strengths. The test is applicable to both textile and steel cord belts and to mechanical and spliced joints. The number of joints in the test loop is decided between the test house and the manufacturer, but at least three joints are run to failure at fixed tensions relevant to the type of belt and joint. The individual and average running time to failure, in hours, is reported.

DIN 22110-3 describes a test to establish the fatigue strength of joints under continuous dynamic pulsating stress. Essentially the test seeks to establish the curve of stress against number of fatigue cycles (the S-N curve) for the belt joints tested. The test loop containing the joint to be tested is positioned over two equally sized pulleys, one of which may be moved to apply stress to the test loop. The diameter of the pulleys depends on the strength of belt tested. The stress is varied in a saw-tooth fashion between limits over a period equivalent to eighteen cycles round the test rig. The lower stress level is fixed at 6.67% of the rated breaking strength of the belt sample, while the higher level is changed from test to test, so that the S-N curve is produced. The value of the stress at which the joint lasts for ten thousand cycles is derived from the S-N curve and used to calculate the relative fatigue strength for the joint in kN/mm width.

The authors are not aware of any work to correlate performance in the UK and German tests.

7.6 Concluding remarks

There are, of course, many other tests that have been developed to assess various aspects of conveyor belt performance and that have not yet been standardised. Many of these are used by manufacturers to help them to assess fitness for purpose for specific applications and also for development work on new products. Some of the tests that Barclay[7.1] developed, flexibility, coefficient of friction, ageing characteristics and water absorption, have been mentioned earlier. However, in addition he did a good deal of work on assessing the resistance of belts to incidental damage by impact, edge rubbing and belt turnover. Some of these tests, such as ageing and water absorption, have not been standardised either nationally or internationally because they do not provide information that is of sufficient value to the user or the manufacturer. Some of the other properties he measured do seem to have been more important, although the test methods have not been standardised. The *Fenner-Dunlop Technical Manual*[7.2] indicates that, in addition to carrying out a wide range of tests to national and international standards, they test for impact resistance, flexing and friction. Clearly these tests provide Fenner-Dunlop with useful information. There may, therefore, be further scope for test standardisation in respect of some of these properties.

7.7 References

7.1 J T Barclay *Conveyor Belting Research. A monograph on the work carried out on conveyor belting at the Mining Research Establishment of the National Coal Board during the period 1950–1966* UK National Coal Board 78pp.

7.2 *Fenner Dunlop Technical Manual* Fenner Dunlop Ltd, Marfleet, Hull, UK.

8
Safety considerations 1 – fire and electrical resistance properties of the belt conveyor

> In the last 20 years, belt conveyors have been responsible for the largest proportion of fires in mines.
>
> UK Deep Mined Coal Advisory Committee, *The Prevention and Control of Fire and Explosion in Mines*

8.1 Introduction

It is, perhaps, surprising quite how dangerous a piece of equipment a conveyor belt installation actually is. The combination of the applied power, the transmission of that power by friction to the belting, the length of any given installation, the numerous rotating parts, and the speed of the belt itself leaves many hazardous circumstances that operators need to be aware of and which they need to guard against. The authors have chosen to divide this key area of conveyor safety into two. In this chapter, safety hazards associated with the fire and electrical resistance properties of the conveyor belting itself are addressed. In Chapter 9 the authors consider those safety issues associated with the interaction of the belt conveyor and people.

Fire associated with the operation of a conveyor is particularly dangerous in any underground mine or tunnel but can be disruptive in surface operations also. This chapter looks at some of the historical events that led to today's testing regimes to establish the fire resistance properties of belt conveyors. Explosions caused by the generation of static electricity on conveyors operating in dusty or gassy atmospheres have proved to be a key safety issue also and this will be commented upon later in this chapter.

8.2 Fire hazards

Because conveyors are driven intentionally by friction, and on occasions subject their moving and static parts to unintentional frictional effects, fire is an ever-present hazard. The results of a fire in a surface installation would normally have only economic effects on the conveyor operator. In November 2006, the

110   Belt conveying of minerals

Queensland Government Department of Mines and Energy website reported the severe damage to a conveyor belt and structure leading to a rail load-out facility on a coal mine surface due to fire breaking out. No injuries were reported but it was stated that the facility was badly damaged. The effect of this incident on production and sales from a mine producing a reported 5.5 m tonnes per annum can only be guessed at.

However, if the application of the conveyor is in an underground tunnel or mine, the effects of a fire are greatly increased due to restricted access and egress for personnel and the potential for death by inhalation of the products of combustion as well as normal heat-related injuries. If one adds to this the additional issue of conveying an inflammable material such as coal and the danger of the possibility of explosions due to the presence of coal dust and methane gas, then the potential danger of a fire associated with a conveyor underground is higher still and its effects increased again.

In their 2004 report the UK Deep Mined Coal Advisory Committee[8.1] state:

> In the last 20 years belt conveyors have been responsible for the largest proportion of fires in mines. Bearing failure, commonly from conveyor idler rollers and drums, causes many fires. In most cases, where idler roller bearings fail, it is because they have been subjected to continual loads far in excess of their specified safe working load.

In the UK the approach to fire prevention on and around belt conveyor installations is multifaceted. The most significant single action is that only fire-resistant conveyor belting is applied in both coal and other underground mines. Additionally, the report mentioned above calls for:

- good engineering design including matching equipment design to duty
- the use of fire-resistant grease in idler bearings
- the use of fire-resistant hydraulic fluids in traction couplings and other hydraulic systems
- the use of brake designs not prone to sticking on
- comprehensive monitoring of wear in mechanical parts
- the minimisation of spillage and dust at transfers
- the provision of means to deal with material removed from the belt surface by belt cleaners
- good clear access to inspect and clean systems and
- the use of fire-resistant materials around conveyor transfer and loading points.

The report stresses the additional problems of conveying a flammable product such as coal and the need to avoid accumulations around moving parts.

The importance of the findings of Committee reports such as that mentioned above can be seen when one encounters industry statistics. Even in 2005, when the UK coal mining industry had shrunk to fewer than ten underground mines, of the eight underground fires reported under statutory procedures to the Health and Safety Executive, seven were the direct result of the operation of conveyors.

8.3 The Cresswell disaster

The stimulus for the approach to fire safety on conveyors in the UK was the disastrous fire that occurred at Cresswell Colliery in 1950, when a belt fire caused by friction of a rubber covered, ply construction, conveyor belt running in the main intake airway at Cresswell Colliery in Derbyshire resulted in the deaths of 80 miners. The miners had been working in districts inbye[1] of, and ventilated by, intake air that flowed over the belt installation that caught fire.[8.2] All of the deceased were reported to have died from carbon monoxide poisoning as a result of smoke and fume inhalation. Attempts to fight the fire were hampered by problems with the water supply in the fire-fighting range, the heat of the fire and the smoke from the fire that reduced visibility in the affected roadways to zero. Additionally, contact with the inbye districts that were affected by fumes from the fire being carried inbye by the ventilation system, was difficult to achieve. No self-rescue apparatus was available to miners in those days, making them particularly vulnerable to the effects of fire in a single intake.

The fire at Cresswell was subsequently found to have been caused, probably, by friction between strips of belting torn from a running conveyor and the remaining running belt. The strips had been torn by a longitudinal cut formed by a piece of rock and had gathered in a delivery chute, where they had been rubbing against the running belt. The fire was found to have spread 558 metres inbye from the source of ignition, before it was isolated from the air supply by sealing the roadway.

Among the comments in the official report of the Inquiry into the fire set up by the UK government[8.2] are the following:

- 'As soon as they are proven in practice and are commercially available, only belts which are non-flammable or are highly resistant to fire should be used [in underground coal mines].'
- 'Automatic safety devices to prevent overloading, belt-slip, overheating, and piling up at transfer points and to detect damaged belts should be developed and used.'
- 'Substantial damage to conveyor belts should be reported forthwith and should be examined by a competent individual who should have authority to stop the conveyor and repair the damage.'

Appendix IV of the report reveals that a major investigative programme on fire hazards in mines was established as a result of this disastrous fire. The report concluded:

> Research confirms that
> (1) the main risk of fire in the type of conveyor belting in ordinary use underground in mines lies in the cotton-duck foundation of the belt;

---

1. Inbye – the direction that leads towards the working face.

112   Belt conveying of minerals

(2) this risk can be reduced by fire-proofing the duck; and
(3) the plastic PVC, if used instead of rubber for the facing of the belt, greatly reduces the risk of fire from the duck.

Conveyor installations using this highly fire-resistant type plastic belting have been on trial underground with promising results, and encouragement is being given by the NCB [National Coal Board] to belt manufacturers to expedite production of this type of belting.

Further conclusions of the Report listed 'important matters to which I [the Chief inspector of Mines at the time] should like to draw attention'. These included the following:

- Roadways (in which conveyor belts run) should be straight, well-graded and large enough in cross-section to allow debris to fall off and not rub against the belt.
- Parts of roadways adjacent to conveyor drives should be of fireproof construction.
- Roadways should be kept free of combustible material, especially coal dust.
- Attendants to be stationed at all transfer points not safeguarded by the systems listed.
- Competent persons to patrol belts whilst they are in operation and for two hours after stopping.
- Effective means for stopping a conveyor from any point along the roadway should be provided.
- Substantial damage to conveyor belts should be reported forthwith, examined and repaired.

All of these steps were put into practice and have formed the basis of underground belt conveying practice, not just in coal mines in the UK, but in all underground mines in the UK, and in many other countries to this day.

8.4   Early research into conveyor fires

Some of the earliest reported work on the potential for conveyors to cause fires is contained in a monograph entitled *Conveyor Belting Research* by J T Barclay.[8.3] This book describes work done by the NCB between 1950 and 1966, firstly on the fire and electrostatic safety of conveyors and then on means to improve their efficiency. It was stimulated by the Cresswell disaster.

Barclay analysed the causes of seventy-five conveyor fires and found the following principal causes and percentage occurrences:

- fires on the belt caused by friction, most commonly when the belt 'stalled' – 32%
- fires in other parts of the conveyor system, mainly idlers failing and setting fire to coal dust or other material – 52%, and

- fires generated outside the conveyor system – 7%.
(9% of fires were unaccounted for).

A belt is said to be stalled when it is stationary relative to the driving pulley or some other rotating component. It is easy to see that in this circumstance frictional heating of the belt could readily occur. The two principal hazards were therefore seen as the generation of fire by frictional heating and the intensification and propagation along the belt of a fire from an external source. To be safe, conveyor belts in coal mines needed to be resistant to both. Barclay's research concentrated first on the generation of frictional heating of and by conveyor belts – the cause of the Cresswell disaster. The work showed that it was very difficult to generate sufficient heat to ignite a belt, or that surface against which it was rubbing, if the belt were moving and impinging on some stationary object (such as a part of the conveyor structure or a wooden pit prop). It was also difficult to ignite a belt that was moving through an external fire unless the fire were very large or the belt moving very slowly. However, with the belt stationary under tension around a rotating drum, then the cotton carcase, rubber covered belts in common use at that time could be ignited fairly readily. Barclay and his co-workers found that the fires originated in the belt carcase after the rubber cover had been worn away, with glowing particles of carcase that burst into flame. Naturally the first thought was to make the carcase fire- and glow-proof by treatment with various chemicals, but all of the methods tried had some drawback, most commonly a reduction in the tensile strength of the carcase. The most fundamental objection to this approach, however, was that it would only have made the belt resistant to fire generated by frictional heating and do nothing to prevent the spread of fire along a belt through the rubber covers and inter-ply layers.

Attempts were made to substitute other materials for the rubber of the covers and inter-ply layers. None was entirely successful. However, in experiments with fabric coated with polyvinyl chloride (PVC) to resist abrasion, it was found that the parts of the fabric immediately beneath the coating did not burn on the application of a test flame. The interior of the carcase did burn. The PVC had melted and impregnated the fabric thus rendering it fire resistant. Suitably plasticised PVC is self-extinguishing and will not propagate flame. Thus it was considered that if the rubber covers and inter-ply layers of a belt were replaced by PVC the belt would be rendered resistant to fire and glow produced by frictional heating and to propagation of fire. Belts of plied construction produced as a result of this somewhat fortuitous finding were found to perform well, both in terms of not propagating fire and of not igniting due to friction. In addition the PVC had no adverse effect on the strength of the carcase. Solid woven belts did not perform so well because the melting of the PVC covers did not penetrate sufficiently far into the innermost parts of the carcase to prevent it glowing due to friction, and it was found necessary to impregnate the carcases of solid woven belts to obtain satisfactory results.

A way forward had therefore been devised in the quest to make conveyor belts resistant to fire and to seek to eliminate incidents involving the tragic loss of life that happened at Cresswell. The route to the production of fully functional and efficient fire-resistant conveyor belts was not without its pitfalls, as will be described later. However the UK coal mining industry had accepted that the need and the means existed to make conveyors much safer. One of the next steps taken was to develop robust methods for the measurement of fire resistance and to set limits on acceptable performance. It is worth dwelling on the way in which this was done in this case because, as we shall see, what was then a relatively straightforward matter, later became much more complicated. This in turn raised fundamental questions about the way in which properties such as fire resistance should be measured.

At an early stage in their investigations Barclay and his co-workers realised that some belts could be resistant to only one of the two fire hazards noted. Some belts would not propagate fire but would produce flame or glow in the frictional rubbing situation, while others were satisfactory in the friction situation but propagated fire readily. One test was thus not sufficient to qualify belts as fire resistant and two tests, each simulating a situation occurring in service, were needed.

8.4.1 Fires caused by friction

For the frictional rubbing situation the test devised was a direct simulation of service in that a length of belt was passed round a horizontal steel drum through 180° and weights were applied to the lower strand to provide tension (Fig. 8.1).

8.1 Drum friction test (former NCB photograph – published by kind permission of Department for Business, Enterprise and Regulatory Reform).

The drum was rotated away from the top strand, i.e., similar to the forward direction of a drive pulley in a conveyor in service. The width of belt sample chosen was six inches (150 mm) as being the best compromise. Wider samples were said to offer no practical advantage while narrower ones tended to break prematurely. The weight applied to the belt sample to provide tension was chosen as 70 lb (31.7 kg) as this weight caused a five-ply cotton carcase rubber covered belt to ignite in one hour. The drum diameter was eight inches (203 mm) and its rotational speed was 190 ± 10 rev/min. The test was to be carried out in still air and in an air current of 500 ft/min (2.54 m/s). The purpose of the air current was to simulate the effect of the mine ventilation. As with many tests, therefore, this one was not a direct reproduction of service conditions, but simulated some elements of it and adopted some parameters on the basis of experimental convenience, mainly driven by the need to complete tests in an acceptable length of time consistent with the need to distinguish between 'good' and 'bad' belts. It was considered essential that the belt should be made to break, i.e., to part on the drum, thus eliminating the frictional hazard. When stronger belts became available, the test regime was modified to allow for the load to be increased incrementally, to give progressively higher tensions. Failure in these tests was indicated by either:

- any sign of flame or glow on any part, or
- a temperature of the surface of the drum exceeding 300 °C.

The temperature limitation was imposed because some belts were found that would produce no flame or glow, but would result in excessive drum temperatures. Coal dust sprinkled on such belts ignited if the drum temperature exceeded 300 °C. This test with some minor modifications is that which is set out in EN 1554:1998.[8.4] Many countries have adopted drum friction tests to qualify conveyor belts for use in hazardous conditions. The vast majority are very similar to that in EN1554:1998, but with some variations in the load levels and the times of application. No country that uses this test allows the sample to ignite. Most do not allow flame at all. Some allow glow to be present, sometimes for defined periods depending upon the category of belt and its potential application. Table 8.1 summarises the test details for the various countries that have this type of test and gives the requirements for acceptance.

8.4.2 Fire propagation

The path to a test for resistance to flame propagation followed a somewhat different route. Large-scale laboratory tests were made to simulate conditions that might occur in service so that a correlation of any proposed test with service conditions could be established. Two types of large-scale test were made:

- Flame tests on full-scale static belt samples in still air and in an air current of 500 ft/min (2.54 m/s)

Table 8.1 Summary of drum friction tests and requirements (by kind permission of Fenner Dunlop Conveyor Belting Europe)

Country & test reference	Width of sample (mm)	Drum diameter (mm)	r.p.m.	Peripheral speed m/s	Peripheral speed ft/min	Area of drum contact in²	Area of drum contact cm²	Tension N	Tension lbf	Tension kgf	Time	Ambient conditions	Max drum temp	Remarks
AUSTRALIA AS4606/AS1332:2000 AS1334.11:1988	150±1	212.8 ±0.3	190	2.12 ± 0.05	417	77	501	343	77	35	To destruction or 2 h	6 tests for equal covers: 3 in still air and 3 in air current (2.0 ± 0.1 m/sec) 8 tests for unequal covers (as in BS 3289)	325°C	Should not exceed max. temp., or no visible flaming or glowing
CANADA CAN/CSA M422-M87 1995	150±2	210±2	195±10	2.2	433	77	495	343 686 1029 1372 1715	77 154 231 308 386	35 70 105 140 175	60 min +30 min +30 min +30 min +30 min	Type A1 4 pieces tested: 2 tests each side in still and moving air	325°C	No flame or glow Belting shall part in 3 h
								343	77	35	3 h	Type A2 4 pieces tested: 2 tests each side in still and moving air	325°C	No flame or glow Belting shall not part during 3 h testing period
								343	77	35	3 h	Type B1-A 2 pieces tested: 1 test each side in still air	400°C	No flame or glow Belting shall part in 3 h
								343	77	35	3 h	Type B1-B 2 pieces tested: 1 test each side in still air	400°C	No flame Glow or sparks allowed Belting shall part in 3 h
								343	77	35	3 h	Type B2 2 pieces tested: 1 test each side in still air	400°C	No flame Glow or sparks allowed Belting shall not part during 3 h testing period
								343	77	35	3 h	Type C 2 pieces tested: 1 test each side in still air	400°C	No flame Glow or sparks allowed

CHINA MT914: 2002	150	210±1	200±5	2.2	433	77	495	343 686 1029 1372 1715	77 154 231 308 386	35 70 105 140 173	60 min +30 min +30 min +30 min To destruction	6 pieces tested: each side in still air and in air current (2.0±0.1 m/s). Worst result repeated	325°C	No flame or glow
EUROPE EN 1554:1998 Method A Still air A1 Fixed end load	150	210±1	200±5	2.2	433	77	495	343	77	35	60 min	4 tests 2 cover side 2 pulley side	325°C	Class A, B2 and C2: no flame and glowing is permissible
A2 Increasing end load								343 686 1029 1372 1715	77 154 231 308 386	35 70 105 140 173	60 min +30 min +30 min +10 min To destruction	4 tests 2 cover side 2 pulley side	or 450°C	Class B1: no flame or glow; Max. temp. <450°C Class C1: no flame or glow; Max. temp. <325°C
Method B Moving air B1 Fixed end load	As A1 above	But in air flow of 2.0± 0.1 m/s												
B2 Increasing end load	As A2 above	But in air flow of 2.0± 0.1 m/s												
EN14973:2006 Class A Class B1 Class B2 Class C1 Class C2												EN1554 Method B2 EN1554 Method B2 EN1554 Method B2 EN1554 Method B2 EN1554 Method B2	No limit 450°C No limit 325°C No limit	No flame allowed No flame or glow No flame allowed No flame or glow No flame allowed

Table 8.1 Continued

Country & test reference	Width of sample (mm)	Drum diameter (mm)	r.p.m.	Peripheral speed m/s	Peripheral speed ft/min	Area of drum contact in²	Area of drum contact cm²	Tension N	Tension lbf	Tension kgf	Time	Ambient conditions	Max drum temp	Remarks
EN12882:2002														
Class 3A, 3B, 4B												EN1554 Method A1	No limit	No flame allowed
Class 5A												EN1554 Method A2	No limit	No flame allowed
Class 5B												EN1554 Method A2	No limit	No flame or glow
Class 5C												EN1554 Method A2	400 °C	No flame or glow
INDIA IS3181:1992	150	210±1	200±5	2.2	433	77	495	343	77	35	60 min	Equal covers, 4 tests, 2 in still air and 2 in air current (2.0±0.1 m/s)	325 °C	Sample fails if there is any flame, glow or max temp is exceeded. For belts with unequal covers, four further tests are conducted, 2 in still air and 2 in moving air using the cover to the drum which gave the worst result in the original 4 tests
								686	154	70	+30 min			
								1029	231	105	+30 min			
								1372	308	140	+10 min			
								1715	386	173	To destruction			
SOUTH AFRICA SABS971:2003	225	450	110±10	2.6	512	280	1591			22.5	15 min	2 pieces tested, 1 test each side in moving air (90±3 m/min)		Must not show any sign of flame or glow
										34	+15 min			
										45.5	+15 min			
										59.5	+15 min			
										75.5	+15 min			
										91.5	+15 min			
										107.5	+15 min			
										123.5	+15 min			

- Tests on an experimental conveyor having 20 feet (6 m) of belting either nine or 12 inches wide (226 or 304.5 mm), run at speeds of up to 600 ft/min (3 m/s). Fires resulting from defective rollers running in coal dust and artificial fires of various descriptions were applied to the belt samples.

The flame test which was developed from this work was intended to test for the tendency of the belt to propagate fire. Essentially it involved the application of a laboratory Bunsen burner to a test piece one inch (25 mm) wide and 12 inches (304.8 mm) long for a period of 30 seconds. Six test pieces were taken in each of the warp and weft directions and since belts sometimes are run with the covers worn off, three test pieces from each direction had the covers removed before being exposed to the test flame. The criteria for passing the test were as follows:

- For the six test pieces with the covers intact the average time for the disappearance of all flame or glow after the removal of the test flame was not to exceed three seconds.
- For the six test pieces with the covers removed the average time for the disappearance of all flame and glow was not to exceed five seconds.

Clearly, Barclay and his co-workers gave some considerable thought to the validity of a test involving such small specimens, for he remarks that a propagation test might be expected to be applied to a full-width sample of belting, but that this would be expensive and inconvenient. Attention was given to the scaling down of the test and experiments were made with test pieces from six inches (150 mm) wide downwards. Test pieces of one inch (25 mm) in width were the most convenient to test and gave consistent results that tied in with the larger scale tests that had been made. Samples larger than one inch (25 mm) wide gave no advantages and increased the exposure time needed for some belts to become fully ignited. Samples of a half-inch (12.7 mm) width gave inconsistent results.

The acceptance criteria were set not only after examination of the correlation with the large-scale tests, but also after consultation between the NCB and the conveyor belt manufacturers. They are said to include a factor of safety because the point at which a belt becomes unsafe depends on the circumstances of its use.

8.5 Questions regarding Barclay's approach

The fire propagation test devised by Barclay is very similar, certainly in principle, to that which is described in ISO 340:2004.[8.5] Both ISO 340 and Barclay's test use 25 mm wide specimens cut from both warp and weft directions, with and without covers and similar burners. However, there are differences in the orientation of the test piece, the time of application of the test flame, the precise type of burner and the acceptance requirements.

The test piece in Barclay's test is horizontal with the burner a fixed distance beneath it. Up to the latest revision of ISO 340 in 2004 there was a choice of specimen orientation relative to the burner, with either the test piece vertical and the burner is held at 45 degrees or the burner vertical and the specimen at 45 degrees. Barclay decided to apply the test flame for 30 seconds because that was sufficient to ignite fully those belts which were flammable. ISO 340 requires the test flame to be applied for 45 seconds. Barclay initially used a Bunsen burner using town gas but later switched to the use of the Barthel or spirit burner, which burns methylated spirit at a specified rate, for the purposes of standardisation. At that time town gas was produced from coal and could be of different calorific value depending on where in the country the gas was made. The NCB at that time employed about 250 000 people and had test facilities in many parts of the UK. It was required by the Mines Inspectorate to carry out fire resistance tests on every batch of conveyor belting purchased. Standardisation was therefore vital. Until the 2004 revision, ISO 340 offered a choice of a Bunsen burner operating on either town gas or liquid petroleum gas, or the Barthel burner. The revision published in 2004 allows only the Bunsen burner with commercial propane gas.

The final difference is the acceptance criteria. ISO 340 does not contain acceptance requirements. However, until relatively recently in some countries if a belt was tested in accordance with ISO 340:1988, it would pass if the total time of afterburn of the six test pieces aggregated to not more than 45 seconds and no single test piece exceeded 15 seconds.

The introduction to the 1988 version of ISO 340, in which there is a choice of three burners and two specimen orientations, states:

> It is stressed that for small scale laboratory tests of the type covered in this International Standard the *correlation of test results with flammability under other conditions is not in any case implied.*

ISO 340: 2004 states right at the beginning of the standard:

> CAUTION – This method of test is *not designed to assess the fire hazard of any given product.* The results may help in the assessment of the *ignition hazard* but should not be used in isolation as evidence that a product or material is safe.

During the process of international standardisation, therefore, it appears that this simple test that Barclay had correlated with large-scale service simulations, changed to one where there is no correlation with flammability under other conditions, the test is not designed to assess the fire hazard and is an ignition test rather than a propagation test.

This does not appear to be progress, so what happened to cause these changes? Did we become more cautious? Are the differences between the two tests so significant? Is there new technology that has had to be taken into account? Is there new knowledge that negates Barclay's finding? The answer is probably a qualified 'yes' to all of these questions.

The NCB was a very large organisation that produced its own standards and had a degree of control over the way in which those standards were used that ISO does not have. In this litigious age it would be amazing if ISO were not substantially more cautious than was the NCB. We do not, at this distance in time, have access to Barclay's results and as far as the authors are aware there has been no study to examine whether ISO 340 correlates with large-scale tests. It would be prudent, therefore, to assume that the differences in rate of heat input to the specimen and the differences in orientation could affect the test results. Belts have become thicker, heavier and stronger over the years and our knowledge of fire testing and fire dynamics has increased enormously.

Tests of the type developed by Barclay and standardised by ISO are used widely to assess fire resistance. Table 8.2 summarises the tests used in different countries and their requirements.

8.6 The European dimension

In 1971 the Mines Safety and Health Commission of the (then) European Community became involved in the issue of conveyor fire safety. The Commission clearly felt under pressure to act to avoid further loss of life in mines through fire. In addition to the belt fire at Cresswell, the hydraulics fire at Marcinelle in Belgium in 1956 had caused the loss of 262 lives. The Commission's working party on 'Rescue Arrangements, Mine Fires and Underground Combustions' examined cases of fires on conveyor belts and found that there were instances where belts had been completely burned despite having been accepted for use underground on the basis of ISO 340. A review of the adequacy of fire safety standards was therefore instituted and a committee of experts was set up to carry this out. The Commission's report,[8.6] published in 1974, contains statistics of mine fires that had taken place between 1 January 1960 and 31 December 1972 in all of the European Community (EC) countries that operated coal mines (Belgium, France, Germany, Netherlands and UK), details of their testing arrangements for conveyor belts and proposals for improved safety testing regimes. The report also states that comparative tests on belts from various countries were carried out.

A total of 472 fires was reported to have occurred in the period, of which 194 had affected the conveyor belt, with some having propagated along the belts. The potential for another Cresswell was therefore substantial.

In France and Germany the only official test prescribed at that time was that arising from ISO 340. However in Germany, new full-scale tests had been developed using samples of up to 50 m length, tested under very realistic conditions. They were also working to develop a new small-scale gallery test. In Belgium, Holland and the UK, the friction test was used. The Belgians and the

Table 8.2 Summary of laboratory scale fire tests and requirements (by kind permission of Fenner Dunlop Conveyor Belting Europe)

Country & test reference	No. & size of test pieces	Allocation of test pieces from sample	Orientation/ cover detail	Time in flame	Air conditions after removal of flame	Aggregate time allowed for flame or glow for each set	Average time allowed for flame or glow for each set	Maximum time allowed for any one test piece	Remarks
AUSTRALIA AS4606/AS1332:2000	10 150×13 mm	5 from warp 5 from weft	covers on	60 s	Still air until flame extinguished then turn on air current @ 1.5±0.15 m/s	Grade S	Grade S	Grade S	Approval test only requires 10 samples
AS1334.10:1994	10 150×13 mm	5 from warp 5 from weft	covers off	60 s		No aggregate time specified	Covers on: 10 s Covers off: 15 s	Covers on: 15 s Covers off: 25 s	Flame only measured Glow not recorded
CANADA CAN/CSA M422-M87 1995	8 150×12.5 mm	4 from warp 4 from weft	covers on covers on	60 s	Still air	Type A1 & A2 Type B1-A, B1-B & B2 Type C	40 s flame 120 s glow 60 s flame 180 s glow 60 s flame No limit on glow	Not applicable Not applicable Not applicable	Burner is Pittsburgh University Bunsen type using technical grade or better methane
CHINA MT914:2002	24 150×25 mm	3 from warp 3 from weft 9 from warp 9 from weft	covers on covers off	30 s 30 s	Still air Still air	18 s 45 s	3 s 5 s	10 s 15 s	Spirit burner is used.
EUROPE EN14973:2006 EN12882:2002 EN ISO 340:2004	6 200×25 mm 6 200×25 mm	3 from warp 3 from weft 3 from warp 3 from weft	covers on covers off	45 s 45 s	Still air Still air	A, B1, B2 & C2: 45 s C1: covers on: 18 s covers off: 30 s	A, B1, B2 & C2: 7.5 s C1: covers on: 3 s covers off: 5 s	15 s C1: covers on: 10 s covers off: 15 s	There must be no appearance or reappearance of flame or glow when test piece is subject to an air flow at 1.5 m/s for 60 s after removal from flame

INDIA IS3181:1992	6 150×25 mm	3 from warp 3 from weft	covers on	30 s	Still air	18 s	3 s	10 s	If one piece flames or glows for over 10 s, but average is within spec., a further 6 pieces are to be tested.
	6 150×25 mm	3 from warp 3 from weft	covers off	30 s	Still air	30 s	5 s	15 s	If one piece flames or glows for over 15 s, but average is within spec., a further 6 pieces are to be tested
SOUTH AFRICA SABS971:2003	6 200×25 mm	3 from warp 3 from weft	covers on	45 s	Still air	45 s	7.5 s	15 s	There must be no appearance or reappearance of flame or glow when test piece is subject to an air flow at 1.5 m/s for 60 s after removal from flame
	6 200×25 mm	3 from warp 3 from weft	covers off	45 s	Still air	45 s	7.5 s	15 s	
USA 30CFR18.65:1978	4 6×0.5 in (152.4×12.7 mm)	2 from warp 2 from weft	covers on	60 s	300 ft/min		1 min flame: 3 min afterglow		Burner is Pittsburgh-University Bunsen type

Dutch in addition did a full scale test in a gallery, in the course of which a piece of belt was submitted to the flame of a standardised propane burner. As we have discussed, in addition to the drum friction test the UK required a test similar to ISO 340, but with more stringent acceptance conditions, to be carried out.

Little commonality was therefore present among the test regimes used for the assessment of fire resistance and the committee of experts of the Mines Safety and Health Commission clearly felt that none was entirely satisfactory. They therefore proposed that the following three criteria be used to determine whether a conveyor belt could be considered fire resistant:

1. In the case of a belt being stalled and the drum continuing to rotate, heating of the belt and the drum should not cause a fire.
2. Exposure of the belt to a small intense fire (caused, for example, by a conveyor roller, the brakes of the drive, rubbing of the belt, or fine coal) does not result in a spread of flame by the belt.
3. In the case of a major fire (burning, for example, oil in a stationary fire, or wood or coal in a propagating fire) the conveyor belt does not burn faster than the ignited material. Beyond the fire zone, the belt should not continue to burn.

While Barclay had identified two fire hazard regimes: generation by friction and propagation of an external fire, the work of the Commission identified three, the second and third being divisions of Barclay's propagation hazard based on the scale of the external fire. On the basis of these criteria they proposed the following testing regimes:

- Where the ISO 340 test alone was used to assess the flammability of conveyor belts; it be replaced by a drum friction test and a propane burner flame test.
- Where tests other than the ISO 340 test were acceptable to individual national mining authorities, the drum friction and propane burner flame test could be used as additions or alternatives to the national tests.

It appears that this second possible testing regime was proposed because the United Kingdom clearly felt that the work of Barclay obviated the need for additional testing.

It was this work by the Commission that discredited ISO 340 as a propagation test and steered the move to larger scale tests for propagation. The propane burner test was adopted in Europe – including the United Kingdom, which built its first fire gallery in 1975 and incorporated the test into British Coal Specification 158 in the 1980 revision and into BS 3289 in the same year. It is now contained in the European Standard EN 12881-1:2005 'Conveyor belts – Fire simulation flammability testing – Part 1: Propane burner tests' as method A.[8.7] The test, originally known as the 'Luxembourg' test and later as the 'two metre' test, takes place in a gallery two metres by two metres in cross-section in

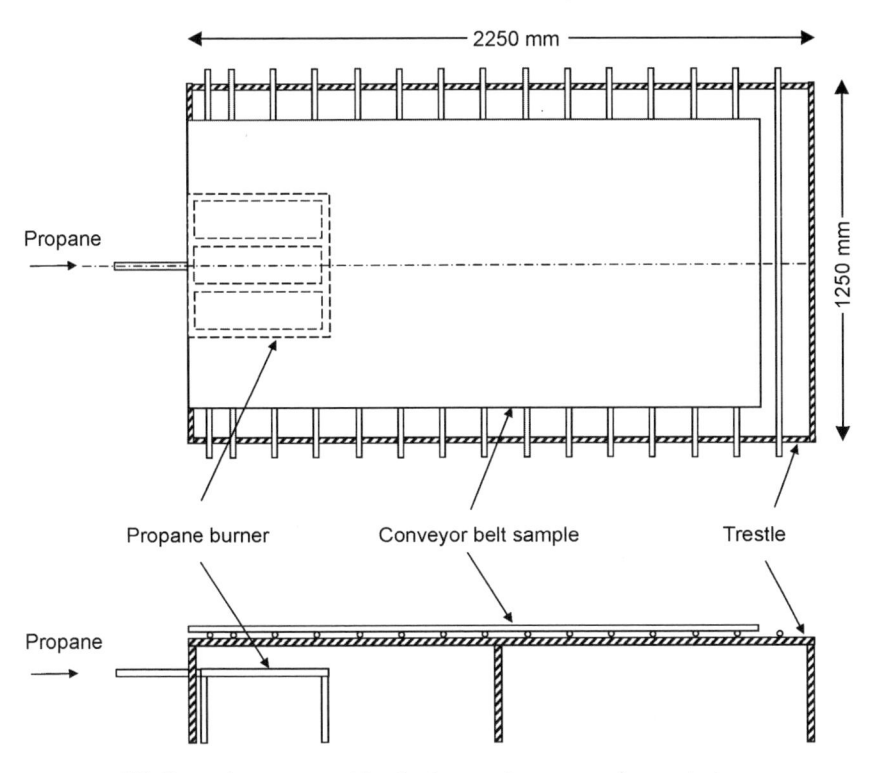

8.2 General arrangement for the two-metre propane burner test.

which a full width belt sample two metres long is laid on a trestle and a specified burner is placed under the leading edge of the belt. The burner consumes 1.3 kg of propane in the ten-minute test period, after the end of which 100 mm of the belt sample must remain undamaged and intact across its full width when all flame and glow have extinguished. The general arrangement for the test is shown in Fig. 8.2. The test procedure and acceptance requirements for the propane burner test differ in certain minor details from those set out by the Commission in its report of 1975. During their adoption of the propane burner test in the period between 1975 and the production of the harmonised European standard EN 12881-1 the countries in Europe had introduced local modifications to the test to cope with what they saw as deficiencies in it. This experience was taken into account in the production of EN 12881-1.

8.7 Safe enough?

During the late 1970s and the 1980s there began to be some concern that belt constructions had become so massive that during the ten-minute burn period some belts were not being fully ignited in the 'two-metre' propane burner test and the test was tending to measure resistance to ignition rather than

126   Belt conveying of minerals

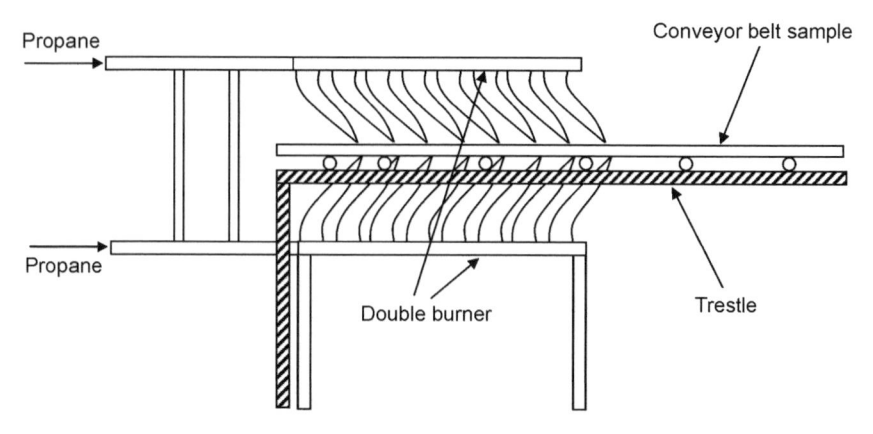

8.3 The Double Burner test.

propagation. Different approaches were taken in Belgium, France, Germany and the UK to this problem and new tests were developed independently from one another in the different countries, as were the acceptance criteria. For steel cord and aramid carcase belts in Belgium and France the test time was increased and the geometry changed to supply heat to both sides of the belt simultaneously. This test is known as the Double Burner (DB) test. The Double Burner test has one burner above and one below the belt sample, which is equidistant between the burners (Fig. 8.3). Belgium and France continued to use the 'Luxembourg' Test for other belt types.

The UK developed the High Energy test which was intended to be applicable to all types of belt. The arrangement for the High Energy test is identical to that of the Luxembourg test, but the length of the test piece, the burner time and the quantity of gas used are greater. As the European Commission report indicated, Germany developed a very large-scale test, applicable to all belt types. This test, the Brandstrecke, involved the burning of 300 kg of wood in a special roadway with an 18-metre long sample of the belt positioned on idlers as it would be in service. The wood was used to line part of the test roadway in what was clearly a direct simulation of a possible service situation. This test arrangement is shown in Fig. 8.4 from which it is clear that the test differs in many significant ways from those which use propane gas as the source of fuel. Germany also applied the 'Luxembourg' Test to all belt types. Details of fuel sources, test piece lengths, gallery dimensions and test times for all of these tests are given in Table 8.3, and the acceptance requirements in Table 8.4.

There is no doubt that belts that would pass any of these tests would possess a high degree of fire resistance and that the situations under which the belts might be used would be substantially safer than if these tests were not imposed. However, the fact that the tests and acceptance requirements in Tables 8.3 and 8.4 differ so widely illustrates the two classic problems associated with trying to

Safety considerations 1   127

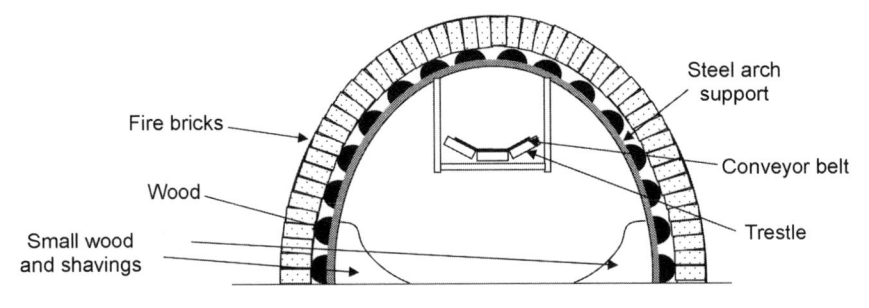

8.4 The Brandstrecke test.

ensure a safe working environment – when is 'safe' safe enough and how do you measure it?

Since conveyor belts are made from organic materials, which can never be completely non-combustible, the approach taken by the European Commission in the 1974 report that the belt should not burn faster than the igniting material, i.e., beyond the burning zone, has merit. Does this imply that in the test the belt must be fully ignited? If it does are we into a cycle of ever-increasing test fire intensities if belts get thicker or more difficult to ignite, or is there a point at

Table 8.3 Comparison of Luxembourg, High Energy, Double Burner and Brandstrecke tests

Test	Luxembourg	High energy	Double burner	Brandstrecke
Source of fuel	Propane gas	Propane gas	Propane gas	300 kg wood
Rate of gas consumption (g/minute)	130	150	260	Not applicable
Exposure period (min)	10	50	20	Not defined
Total gas consumption (kg)	1.3	7.5	5.2	Not applicable
Test piece length (m)	2	4	2.5	18
Test piece width (mm)	900 or 1050	900 or 1050	1250 or full width	Full width
Air speed (m/s)	1.5	1.5	1.5	1.2
Arrangement of burner(s)	Single burner below belt	Single burner below belt	Two burners; one above belt and one below	Wood lines roadway over distance of 3 m
Gallery cross-section	2 × 2 m	2 × 2 m	2 × 2 m	3.5 × 2.9 m

Table 8.4 Acceptance criteria for Luxembourg, High Energy, Double Burner and Brandstrecke tests

Test	Acceptance criteria
Luxembourg	Piece of belt left undamaged over full width of sample
Double burner	Piece of belt left undamaged over full width of sample
High energy	1. Length of belting undamaged over full width of sample shall exceed 2250 mm, or 2. The maximum average temperature rise shall not exceed 90 °C and the length of belting consumed shall not exceed 2000 mm and length undamaged shall exceed 250 mm, or 3. The maximum average temperature rise shall not exceed 80 °C and the length of belting consumed shall not exceed 2250 mm and length undamaged shall exceed 250 mm.
Brandstrecke	Propagation to extend not more than 10 metres past fire source

which we have to say that the test fire is unrealistically large or of unrealistic duration? If a belt can resist ignition during exposure for 50 minutes to a 100 kW burner, as in the High Energy test, is it not sufficiently fire resistant for practical purposes? Experience in the UK coal industry suggests that the answer to this question is 'yes', but if the criterion is that the belt has to be fully ignited then it is 'no'. In the authors' experience there is no doubt that the fire source in the Brandstrecke test is sufficiently large to ignite any belt in current production, or likely to be made in the foreseeable future. Yet a belt which passed that test failed the High Energy test, even though it was fully ignited, because of the difference in acceptance requirements. We are also aware that the High Energy test does not fully ignite all belt types and have carried out a Double Burner test on a 2625 kN/m (Type 15) textile carcase belt that passed the High Energy test. It burned out completely in the Double Burner test.

Despite the problems with large-scale gallery tests identified above, they are widely specified and used. Table 8.5 provides a summary of such tests and their specifications.

8.8 Australian studies

While the higher energy tests were being developed within Europe, work was also being done in Australia on fire safety testing of conveyor belts for use in coal mines. At the Londonderry Industrial Safety Centre, Green and Bilger[8.8] critically reviewed the (then) current fire tests and the knowledge of fire processes, and identified alternative approaches to assessing fire hazards. They considered that the traditional approach to material testing suffers from three significant deficiencies:

Table 8.5 Large scale gallery tests used throughout the world with acceptance requirements (by kind permission of Fenner Dunlop Conveyor Belting Europe)

Country & test reference	Dimension of gallery	Air speed (m/s)	Sample size	No. of samples	Test time	Fuel Type	Fuel Consumption	Pressure (MPa)	Conditions of acceptance
AUSTRALIA AS4606/AS1332:2000 AS1334.12:1996	2.4 m high 2.5 m wide	1.5±0.1	Length 2 m Width: 1050–1200 mm	2 for belts of equal cover thickness 3 for belts of unequal cover thickness	Remove burner after 10 min	Propane	1.3±0.05 kg	0.2	The length remaining undamaged over the full width shall exceed 250 mm
CANADA CAN/CSA M422-M87 1995	2 m high 2 m wide	1.5±0.2	Length: 4 m Width: 900 mm	2 Conduct testing on both sides	Type A1/A2 0.1 min for each 0.1 mm of belt thickness (10 min minimum) Type B1-A, B1-B & B2: 10 min	Propane	130±5 g/min		A full width piece shall remain substantially undamaged. If any test is terminated prematurely due to the exhaust gases exceeding 200 °C, the belting shall be deemed to have failed the test
CHINA MT914:2002 A. Normal gallery	2 m high 2 m wide	1.5±0.1	<1250 S, 800 mm wide >1250 S, 1000 mm wide 2 m long	2 Conduct testing on both sides	Remove burner after 10 min	Propane	1.3±0.05 kg	0.16	The length remaining undamaged over the full width shall exceed 250 mm
B. Higher energy			<1250 S, 800 mm wide >1250 S, 1000 mm wide 4 m long		Remove burner after 50 min		7.5±0.25 kg	0.2	The length remaining undamaged over the full width shall exceed 2250 mm

Table 8.5 Continued

Country & test reference	Dimension of gallery	Air speed (m/s)	Sample size	No. of samples	Test conditions Test time	Test conditions Fuel Type	Test conditions Fuel Consumption	Test conditions Fuel Pressure (MPa)	Conditions of acceptance
EUROPE EN12882:2002 EN14973:2006 EN12881-1:2005 A: single burner	1.9–2.25 m high ×	1.5±0.1	Width: 1.2 m Length: 2 m	A&B: 2 (Conduct testing on both sides)	A: Remove burner after 10 min	Propane	A: (130±5) g/min	0.16	A: Length (remaining undamaged over the full width) > 100 mm
B: Double Burners	1.9–2.75 m wide	1.5±0.1	Width: 1.2 m Length: 2.5 m		B: Remove burner after 20 min		B: (130±5) g/min	0.16	B: Some undamaged belt left
C: mid scale	BELT cabinet	1.0±0.05	Width: 230 mm Length: 1.5 m	C: 2 pieces for belts of equal cover thickness 3 pieces for belts of unequal cover thickness (3rd test conducted on side down giving worst result)	C: Remove burner after 50 min		C: (11.3±0.2) g/min		C: Length (remaining undamaged over the full width) > 600 mm or length consumed (by mass) < 1250 mm and max. temp. <140°C
INDIA IS3181:1992	2 m high 2 m wide	1.5±0.1	Length: 2 m Width: 900 mm for belts up to and incl. 1140 kN/m 1050 mm for belts >1140 kN/m	2 pieces for belts of equal cover thickness 3 pieces for belts of unequal cover thickness (3rd test conducted on side down giving worst result)	Remove burner after 10 min	Propane	1.3±0.05 kg	0.158	When all combustion has ceased, there must be a minimum of 250 mm remaining undamaged across the full width

- tests must simulate the materials, configuration and fire environment
- scaling effects are a serious problem
- a proliferation of tests to simulate all important configurations and fire environments is needed.

They believed that the study of the various phases of the development of a fire could lead to the determination of critical events in the process and thus to improvements in the control of fires. Experiments were needed that separate the processes taking place during a fire and measure materials characteristics at each phase to produce a model of the whole fire process. It was assumed that the necessary measurements could be made in small-scale tests and that the results would be dependent only on the material rather than the test method.

Working along similar lines, Quintiere[8.9] discussed the application of flame spread theory to identify suitable parameters to characterise the fire performance of materials. He also argued that current fire performance tests give results that are specific to a given situation and should not be applied to predict behaviour in other situations. In this study, horizontal flame spread under natural convection conditions was considered for many different materials. Equations were developed which related the velocity of flame spread to certain materials characteristics (thermal conductivity, heat capacity, density, ignition temperature), heat input from the flame and heat input from the environment. The materials parameters identified were the ignition temperature. T_{ig} and the product of thermal conductivity, density and heat capacity, $k\rho c$. Given these and a measure of the flame heat transfer under appropriate conditions, flame spread could be predicted.

This approach was used by Apte et al.[8.10] in the study of flame spread over horizontal polymeric surfaces, the findings from which were intended to be applied to conveyor belts for use in mines. Good agreement was achieved in the trends of flame spread velocity predicted from Quintiere's equations and those from actual measurements, but correct values of the effective materials parameters T_{ig} and $k\rho c$ were needed for the formulae to be tested adequately. The work was carried further in an extensive project[8.11] aimed at using small-scale tests – as per the approach of Green and Bilger above – together with predictive methods to eliminate the need for large-scale gallery tests on conveyor belts intended for use in coal mines in Australia. In this study the large-scale gallery tests were carried out basically to the Luxembourg method (ten-minute burn), while the Cone Calorimeter and the Factory Mutual Research Corporation tests were made on small samples.

The Cone Calorimeter test is described in the paper by Babrauskas and Parker.[8.12] A sample of material approximately 100 mm square is subjected to a radiant heat flux that may be controlled to various levels. The sample may be positioned horizontally or vertically. A spark igniter is used to ignite the pyrolysis vapours. The test measures the ease of ignition, heat release rate and pyrolysis mass loss, and analyses the combustion products.

The Factory Mutual Research Corporation (FMRC) test can be set up for ignition or propagation tests. The heat release rate is calculated from the oxygen depletion or from the concentration of carbon dioxide and carbon monoxide in the exhaust gases. The sample burning rate is measured using a load cell. For the ignition test, a horizontal 100 mm square sample is subjected to a radiant heat flux. A pilot flame is used as an ignition source. The ignition delay time is measured for a range of radiant heat fluxes. In the propagation test the bottom 150 mm of a vertically positioned 100 mm × 600 mm sample is exposed to a radiant flux of 50 kW/m². A pilot flame is used for ignition. An upward stream of oxygen-enriched air flows over the sample. The exhaust gases are analysed.

The parameters measured in the gallery and FMRC tests are different measures of the flame spread rate. The terms in Quintiere's equation were manipulated to derive:

- the thermal response parameter (TRP), which is a function of $k\rho c$, and depends only on the material properties
- the heat release rate (HRR), which is a function of the heat contribution from the flame, and
- the flame propagation index (FPI), which is related to the velocity of flame propagation and is a function of both HRR and TRP.

A FPI was defined for the FMRC test based on an empirical correlation.

Thirty-one belts were tested in the Cone Calorimeter and Gallery tests, and of these, twenty were also tested in the FMRC test. HRR, TRP and FPI values were determined for all three types of test and correlations between the tests were examined. These correlations were good for those belts that convincingly passed or failed, but less good for marginal cases. Some belts passed one test but failed the other. The authors concluded that a full replacement of the Gallery test by an FMRC or Cone Calorimeter test was not justified by their findings. The authors further concluded that the acceptance criterion for the Luxembourg test could be misleading since it only tested for the extent of fire damage on the surface of the belt. The compliance criteria should also include heat release rate and various parameters associated with the continued integrity of the sample during the test. This approach was, in fact, taken in the determination of the criteria for acceptance in the High Energy Propane Burner test by the specification of maximum allowable exhaust gas temperature rises and length destroyed by weight loss.

This work, while very useful, was not completely successful in producing a predictive model for fire propagation based on the results of small-scale tests. Although not directly relevant to the present discussion it is worth noting that other workers[8,13] have used results generated in the cone calorimeter in a predictive model for performance of building materials in a large-scale test with some success.

8.9 Mid-scale galleries

An approach that has not so far been discussed is the use of the so-called mid-scale tests. These tests involve laboratory-sized fire galleries that are essentially miniature versions of galleries such as that required for the Luxembourg test. They are distinguished from small-scale tests by the size of sample used in them and are intended to measure propagation of a fire along a belt rather than some more fundamental property, such as heat release rate, although it may be possible to infer such properties from data obtained in mid-scale tests. Two such galleries have been used, one defined in DIN 22 118[8.14] and the MSHA gallery developed by the US Mine Safety and Health Administration to carry out their 'BELT' (Belt Evaluation Laboratory Test) test. The DIN gallery is 2500 mm long with an opening measuring 350 × 350 mm, whereas the MSHA gallery is 1520 mm long and 460 mm square in cross-section. The flames from the burner in the MSHA impinge on the end of the sample rather than underneath as with the DIN and Luxembourg arrangements. Mintz[8.15] carried out a comparison of performance in the Luxembourg-type gallery and the MSHA mid-scale gallery. He found generally good agreement between the two but also found that the MSHA gallery could burn out almost any belt if the burner application time were increased sufficiently. He remarked that the most useful method of ranking relative flammabilities of belts was through the total heat input, which basically equates to the burner application time.

The present authors were involved in the use of the mid-scale MSHA test apparatus as part of a study into alternative test arrangements that would enable the remaining mines to maintain the level of safety afforded by the High Energy test when economic necessity, caused by the closure of most of the UK coal mining industry, forced the shutting of the UK's large scale fire testing facility.[8.16,8.17] The work involved characterising the response of the large gallery to different heat inputs both without and with belt samples, using thermocouples to measure exhaust gas temperature rises, anemometers to measure air speeds, oxygen depletion measurements and thermocouples attached to the belt samples to measure temperature rises and flame front velocities. Figure 8.5 shows a fire test being carried out on a belt in the large gallery. It was then sought to reproduce the burning behaviour of a range of conveyor belt samples in the MSHA gallery by adjustment of test arrangements (air flow rate, burner and trestle geometry, rates of heat input), again measuring exhaust gas temperatures, oxygen depletions and flame front velocities. Figure 8.6 shows the modified test arrangement in the MSHA gallery. The work was successful in its primary objective of providing a test that will reproduce the burning behaviour and performance of the large gallery. It also produced the interesting finding that by varying the test conditions in the MSHA gallery, it is possible to rank belts in a different order to that obtained in the large-scale gallery.

Considerable experience with the test developed has now been accrued and it

134   Belt conveying of minerals

8.5 Fire test in large-scale gallery. Crown Copyright, provided courtesy of Health & Safety Laboratory (HSL).

has been incorporated into EN 12881-1[8.7] and requirements for passing the test into EN 12882 'Conveyor belts for general purpose use – Electrical and flammability safety requirements'[8.18] and EN 14973 'Conveyor belts for use in underground installations – Electrical and flammability safety requirements'.[8.19] These standards were produced to assist with the European Union's drive to

8.6 Modified test arrangement in MSHA gallery.

remove barriers to trade. When discussions started among the countries of Europe, the variety of fire tests and acceptance requirements then in force led to some 'full and frank' discussions within the standardisation committee dealing with conveyor belts. All countries believed that the tests they were using gave them safe situations and believed that their records proved it. Naturally enough they were reluctant to change to tests with which they were not familiar, because of the perceived risk to personnel. This experience tends to suggest that in reality there is a substantial factor of safety built into all of the test requirements.

EN 12882 and EN 14973 use hazard identification and risk assessment principles to categorise circumstances of use and require appropriate test and acceptance requirements. They also allow for the presence of other methods of ensuring a safe working situation, the so-called secondary safety devices such as belt slip detectors or water deluges. The properties of the conveyor belt itself constitute the primary safety device. Thus, for example, where there is no flammable atmosphere present, glow is allowable in the drum friction test and there are no temperature limitations on the drum surface. However, where there is a flammable atmosphere potentially present and there are combustible dusts and no secondary safety devices, there must be no flame or glow and the drum temperature must not exceed 325 °C. The introduction of risk assessment is no doubt a major step forward in accident prevention, but for the supplier or user can be much more daunting than the simple requirement to comply with a given standard. In essence both Barclay and the European Commission identified the hazards and quantified the risks associated with fires on conveyors.

The following four European standards relate to the propagation of fire on conveyor belts:

- EN 12882 'Conveyor belts for general purpose use – Electrical and flammability safety requirements'
- EN 14973 'Conveyor belts for use in underground installations – Electrical and flammability safety requirements'
- EN 12881-1 Conveyor belts – Fire simulation flammability safety testing – Part 1: Propane burner tests
- EN 12881-2 Conveyor belts – Fire simulation flammability safety testing – Part 2: Large-scale test.

8.10 Concluding remarks on conveyor fire safety

So where does this leave us in terms of fire safety tests for conveyor belts? Barclay[8.3] sought to use a small-scale test for fire propagation because larger scale tests would be costly and inconvenient. To this end he carried out tests to correlate small- and large-scale performance. The European Commission[8.6] drove the move towards large scale gallery tests but other workers[8.8,8.9,8.10] wished to move back to small-scale testing, not only because large-scale tests were expensive and inconvenient, but particularly because they could not be used as predictive tools.

In recent years concerns have also been raised that large-scale tests can be polluting to the environment. Whilst fire resistant conveyor belts do go a long way towards removing the fire hazard in mines and preventing a repeat of the dreadful events at Cresswell referred to earlier, any lay person who has an opportunity to observe a full-scale gallery test in operation will be amazed at the large quantities of dense black smoke that emerge from a fire-resistant belt under test. Should this smoke be experienced in an underground environment, then its presence would be no small issue for the miners affected. In the fire test gallery, emission of this smoke into the atmosphere has, in the experience of the authors, caused 'problems with the neighbours'. For this and other economic reasons, there can be little doubt that the use of large-scale fire gallery testing has a limited future and that the way forward lies with small-scale tests. Until conveyor belts can be made from materials that are completely non-combustible, however, there will always be circumstances in which they will burn and unless conveyor technology changes radically the hazards identified in EN 12882 and EN 14973 will remain. However, due to the fact that belts are available that will meet the present testing regimes for belting for underground use, the risk of belts causing or propagating fires should be very much reduced. However, this is not necessarily so.

The most recent mine conveyor fire that resulted in fatalities, of which the authors are aware, was the fire at the Aracoma Coal Company's Alma No. 1 mine in West Virginia in January 2006.[8.20] This fire resulted in the death of two miners who were overcome by carbon monoxide from the products of combustion whilst attempting to escape from the mine. A gate road conveyor belt caught fire in a storage unit close to its main belt drive, probably due to collapsed bearings and subsequent running friction due to misalignment of the belt. Due to the system of ventilation in the multi-entry mining layout applied at this mine, as in much of the United States, the belt did not run in the major intake airway. Therefore smoke and other fumes were not immediately identified and when they were, the longwall team on the affected district was able to escape into intake air and out of the mine safely.

However, an inbye development team in another district, when they were eventually contacted about the fire, ran into thick smoke from the fire when exiting the mine. Two of this team became separated from their colleagues and died from smoke inhalation. The water sprinkler system at the belt drive was reported to be switched off and did not activate. Attempts to fight the fire were hampered by problems with the fire-fighting water supply.

The similarities between this fire and the Cresswell disaster are chilling. There is a danger that the loss of large-scale facilities, the decline of some major industrial bases and economic pressures may lead to the erosion of the factor of safety built into current belting technology. However, the greatest danger to those working underground lies in forgetting or being unaware of the lessons of the past. The fact that whilst conveyors are applied underground, such events are possible even in modern times, means that all attempts at reducing the possibility of fire should be embraced by mine or tunnel operators. It should be noted, however, that as a result of the Aracoma fire, the authorities in the USA are undertaking a comprehensive review of their strategy regarding conveyor fires in mines.

8.11 Electrostatic hazards

It is well known that static electricity can be produced on conveyor belts running round a system of pulleys. Occasional instances of the phenomenon occurring in coal mines were reported by Barclay[8.3] on some of the first fire-resistant belts. In every case reported, the visual evidence of the build-up of static was by corona discharges between the belt and idlers as the metal fasteners joining belt lengths together approached the idlers. In coal mines the principal hazard associated with such discharges is that they could trigger explosions of methane gas or coal dust. However, coal mines are not the only circumstance where this hazard exists. Dust explosions are well-recognised and well-documented hazards in many industries. The food processing industry handles and sometimes creates finely divided organic materials, which, when dispersed in air, form a dust cloud having a composition within the explosive range which can readily be ignited to burn with explosive violence. An explosion, typical of these circumstances occurred at General Foods Ltd in Banbury[8.21] UK in 1981 and was the subject of a report issued by the HM Factory Inspectorate. In this particular incident, a violent explosion was caused when powder handling equipment malfunctioned. The subsequent investigation noted that 'once the dust cloud was produced there were three possible sources of ignition'. One of these sources was 'an incendive spark arcing from the electrostatic charge generated by the flow of material through the system'.

Although this was not deemed the cause of the explosion in this particular case, it does show the importance of the potential for dust-laden air, both in an underground coal mine or in a surface application where dust is prevalent to be similarly ignited by an electrostatic discharge from a non-metallic product such as a conveyor belt.

In general when two dissimilar materials are brought together and then separated there will be a mutual charging due to electron or ion transfer between them. A conveyor belt passing round a pulley or roller is just such a case of dissimilar materials being brought together and separated. Because of the nature of the charging mechanism, earthing of the rollers or the conveyor will not prevent the initial charging. The build up of charge is balanced by a leakage of charge back to the roller. If the leakage is high then relatively little charge will be left on the belt. Control of the leakage resistance of the conveyor belt

Table 8.6 Tests for electrical resistance of conveyor belts (by kind permission of Fenner Dunlop Conveyor Belting Europe)

Country & test reference	No. of tests	Sample size	Current applied to electrodes	Sample conditions	Electrode dimensions	Composition of electrodes	Contact agent	Base sheet	Conditions of acceptance
AUSTRALIA AS4606/AS1332:2000 AS1334.9:1982	4 (2/side)	No less than 300 × 300 mm	40 to 1000 V d.c. energy loss < 1 W	>2 hrs at ≤23°C and 70% RH	2 electrodes: a. cylinder 25 mm dia. × 32 mm high; b. Annular ring: Internal dia.: 125 mm; External dia.: 150 mm; Height: 22 mm	Brass + Additional tinfoil	Anhydrous polyethylene glycol, soft soap and water	An insulating sheet a little larger than sample and no less than 1.5 mm thick	≤300 MΩ
CANADA CAN/CSA M422-M87:1995	4 (2/side)	300 × 300 mm	energy loss < 1 W	>2 hrs at 20±2°C and 50±5% RH	As above	Brass	As above	An insulating sheet a little larger than sample	≤300 MΩ
CHINA MT914:2002	6 (3/side)	No less than 300 × 300 mm	50 to 500 V d.c. energy loss < 1 W	>24 hrs at 23±2°C and 65±2% RH	As above	Brass	As above	As above	≤300 MΩ

Standard	Electrodes	Sample size	Voltage	Conditioning	Electrode material		Insulating support	Requirement
EUROPE EN14973:2006 EN12882:2002 EN ISO 284:2003	4 (2/side)	No less than 300 × 300 mm	40 to 1000 V d.c. energy loss < 1 W	> 2 h 23±2°C and 50±5% RH; For belt with textile carcass: 20±2°C and 65±5% RH In tropical condition: 27±2°C and 65±5% RH	Brass	As above	As above	≤300 MΩ
INDIA IS3181:992	4 (2/side)	No less than 300 × 300 mm	40 to 1000 V d.c. energy loss < 1 W	> 2 h at 27±2°C and 65±5% RH	Brass Note: additional tin foil electrodes may be used if surface is insufficiently smooth	As above	Polyethylene sheet > 2 mm thick and 300 × 300 mm	≤300 MΩ
SOUTH AFRICA SABS971:2003	4 (2/side)	No less than 300 × 300 mm	40 to 1000 V d.c. energy loss < 1 W	As in Europe	Brass	As above	An insulating sheet a little larger than sample	≤300 MΩ

therefore provides a method of controlling the possible hazard arising from static electricity on conveyors. Barclay reports on early experiments in which measurements of surface resistance of belts were made using the method which has now been universally adopted for this purpose. He also ran belts having surface resistance values between 10^8 and 10^{11} ohms on a full-scale experimental conveyor on which the charge on the conveyor could be measured. He found that for belts having surface resistance values of less than 10^9 ohms no charge was retained on the conveyor, whereas for belts with resistances of 6×10^9 ohms and over charge was retained and under some circumstances discharges at the joints were observed. The value of 3×10^8 ohms for the surface resistance of conveyor belts that was established includes a factor of safety to allow for inconsistencies in processing. This value has been universally adopted for conveyor belts for use underground. The method of measurement, which is that proposed by Barclay, is given in ISO 284[8.22] and the value of 3×10^8 ohms is specified in EN 14973.[8.19] Table 8.6 outlines the tests for electrical resistance used by various countries and their acceptance requirements, and gives their specification references.

8.12 References

8.1 UK Deep Mined Coal Advisory Committee *The Prevention and Control of Fire and Explosion in Mines* Health and Safety Executive 2004.
8.2 'Accident at Cresswell Colliery, Derbyshire' Report by Sir Andrew Bryan The Ministry of Fuel and Power, June, 1952 HMSO London.
8.3 J T Barclay 'Conveyor Belting Research A monograph on the work carried out on conveyor belting at the Mining Research Establishment of the National Coal Board during the period 1950–1966' UK National Coal Board 78pp.
8.4 EN 1554:1998 'Conveyor belts – Drum friction testing'.
8.5 ISO 340:2004 'Conveyor belts – Flame retardation – Specifications and test method'.
8.6 First report on tests and criteria of flammability of conveyor belts with fabric core used in mines of coal in the European Community countries Mines Safety and Health Commission Luxembourg 1974.
8.7 EN 12881-1:2005 'Conveyor belts – Fire simulation flammability testing – Part 1: Propane burner tests'.
8.8 Green A R and Bilger R W, *A Review of Flammability Test Methods for Evaluating the Fire Resistance of Materials used Underground*, Londonderry Industrial Safety Centre, Australia, 1984.
8.9 Quintiere J G, 1988, 'The application of flame spread theory to predict materials performance', *J. Research National Bureau Standards*, 93(1), 61–70.
8.10 Apte V B, Bilger R W, Green A R and Quintiere J G, 'Wind-aided turbulent flame spread and burning over large-scale horizontal PMMA surfaces', *Combustion and Flame* 85: 169–184, 1991.
8.11 Apte V B, Bilger R W, Tewarson A, Browning G J, Pearson R D and Fidler A, 'End of Grant Report on ACARP Funded Project C5033 – An Evaluation Of Flammability Test Methods For Conveyor Belts', December 1997.

8.12 Babrauskas V and Parker W J, 'Ignitability measurements with the cone calorimeter' *Fire and Materials* 11, 31–34, 1987.

8.13 Goransson U and Wickstrom U, 'Flame spread predictions in the room/corner test based on the cone calorimeter' *Fire and Materials* 16, 15-22, 1992.

8.14 DIN 22 118 1991 'Conveyor Belts With Textile Plies For Use In Coal Mines: Fire Testing'.

8.15 Mintz K J, 'Evaluation of laboratory gallery fire tests on conveyor belting' *Fire and Materials* 19, 19-27, 1995.

8.16 Yardley E D and Stace L R, 'Fire Safety Testing of Conveyor Belts' Report of work carried out under HSE RSU Contract Reference 416/R04.

8.17 Yardley E D, Williams M, Wymark S and Stace L R, 'Development of a small-scale fire propagation test for conveyor belts' *Mining Technology (Trans. Inst. Min. Metall. A)* March 2004 Vol. 113 A73.

8.18 EN 12882:2002 'Conveyor belts for general purpose use – Electrical and flammability safety requirements'.

8.19 EN 14973:2006 'Conveyor belts for use in underground situations – Electrical and flammability safety requirements'.

8.20 West Virginia Office of Miner's Health, Safety and Training 'Alma No. 1 Mine Fatal Investigation Report' September 2006.

8.21 Health and Safety Executive *Corn Starch Dust Explosion at General Foods Ltd Banbury, Oxfordshire 18 November 1981* HMSO 1983.

8.22 ISO 284:2003 'Conveyor belts – Electrical conductivity – Specification and test method'.

9

Safety considerations 2 – nip points, stored tension, man-riding and materials transportation on belts

> ... the speed of the belt conveyor is such that conscious thought is not possible and in less than 0.5 second, tool, hand and arm are drawn into the nip point.
>
> M. Hollyoak, Seminar on Bulk Handling Conveyors

9.1 Introduction

This second chapter on safety issues associated with the operation of belt conveyors deals with those hazards posed by the belt conveyor when there is an interface between people and the machine. The issues discussed are broadly grouped under four headings. These are:

1. accidents associated with personnel being drawn into nip-points
2. the hazard represented by stored tension in a stationary belt
3. the safe operation of belt conveyors for man-riding and
4. the rules governing the safe transport of working materials on belt conveyors.

9.2 Nip point accidents

The first hazard that is considered is injuries (or death) caused by personnel being drawn into places where the moving belt meets a rotating element. These places in a belt system are known almost universally as 'nip points' (Fig. 9.1) and represent the most unpleasant, most commonplace and most avoidable of all conveyor related accidents. The terminology does not do justice to the hazard, suggesting as it does the infliction, perhaps, of a blood blister rather than the loss of a limb.

The authors have gathered accident details from a number of countries taken almost at random covering a wide spread of both time and distance. What is striking is the repetitive nature of accidents in which people are drawn into nip points. Some examples of reports on these accidents are given below.

From Martin Engineering's website in the United States, quoting an MSHA report for all mines during the period 1996–2000, comes the following quotation:

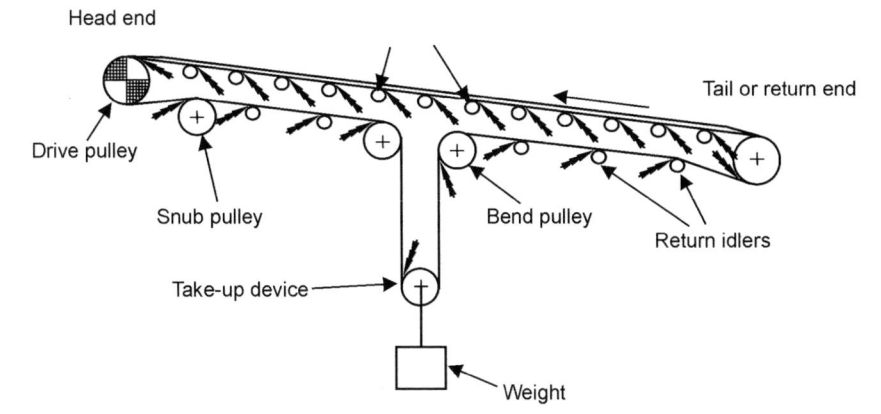

9.1 Diagram showing nip-points in a belt conveyor.

> The most commonly reported cause of accidents around conveyors is getting caught by the moving conveyor belt or pulley. This accounted for 290 of the 459 injuries and 10 out of 13 fatalities.

The report proceeded to list the activity being performed at the time of the accidents. These included working near poorly guarded equipment, using a hand or tool to remove material from a moving conveyor, attempting to remove or install guards on a moving conveyor, and getting clothes or hair caught. It stressed the need for training of personnel as the key preventative measure. The second measure was the need to ensure the correct alignment and working condition of the conveyor to prevent spillage that subsequently needs to be cleaned up and that requires employees to work near to belt conveyors whilst they are in motion.

From Workplace Standards Tasmania in Australia in 2000 came the following report:

> A worker sustained serious injuries when dragged into a moving conveyor whilst carrying out cleaning duties ... The worker was sweeping beneath the conveyor when the handle of the broom was 'nipped' by the moving conveyor belting which in turn dragged the worker's left arm between moving belt and return idler pulley.

The New South Wales (Australia) Department of Primary Industries, in its November 2006 report, describes similar circumstances:

> The operator of a crushing and screening plant at a quarry sustained major injuries and amputation of two and a half fingers. ... He attempted to free a blockage in a transfer chute, miscued and jammed the bar in the gap between the head drum of the conveyor feeding the chute. ... The conveyor pulled the bar and the operator's glove in to the gap.

From the Chief Inspector of Mines of South Africa, a March 2002 report states the following:

144   Belt conveying of minerals

Eight fatal conveyor accidents were reported during 2001 and 66% of these accidents occurred during spillage cleaning operations at the tail pulley.

Finally, from a report in Massachusetts in the United States, published in 1994, comes the following:

> A labourer died from injuries sustained when his left arm became caught between the belt and pulley of a conveyor system ... the victim was working alone removing fallen debris from the conveyor frame at the time of the incident.

Of course, similar accidents have been reported in mines and quarries in the UK, where conveyor belts have been commonplace for many years. The Health and Safety Executive (HSE) website reported two fatal accidents where, in one a conveyor cleaner had become trapped in a nip point when he had crawled past side guards to access the underside of the conveyor to clean spillage. In the second, an attempt to clean compacted fines from a rotating deflecting roller had resulted in the man being drawn by his arms through a 125 mm gap between guard and belt.

A Mines Inspector working for the UK Health and Safety Executive[9.1] noted that if a rake or shovel is drawn into a nip point, the speed of the belt conveyor is such that conscious thought is not possible and in less than 0.5 second, tool, hand and arm are drawn into the nip point. The CEMA conveyor book published in the USA gives a very similar example.[9.2]

The only way to avoid constant repetitive incidents such as those listed above is to keep people entirely away from the nip point whilst the conveyor is in motion. The most common method of achieving this is by the guarding of nip points (Fig. 9.2). However, this has to be combined with training of personnel

9.2 Guarding of nip points (by courtesy of ATH Resources Ltd).

who frequently have to work alone and who on many occasions have been known to remove guarding to 'make their maintenance task easier' whilst not stopping the belt, together with provisions to remotely stop and lock off the belt. High visibility signage is also common practice as a constant reminder to personnel.

Guards are specified for almost all situations in which the moving belt comes into contact with a roller, with the notable exception of normal run of belt top and bottom idlers. The HSE on their UK website note that whilst these points are also nip points, the risks of serious injury from them are quite low and, in fact, there is no history of accidents. HSE do supply a note of caution here. The increased power transmitted by modern belts, the higher loads and stronger, thicker belting, and the increased size of idler rollers do lead to the need to assess normal custom and practice with respect to guarding and access. Access being gained underneath a running conveyor, possibly in a situation where it is normal for people to regularly pass beneath that conveyor, can also lead to potential problems and calls for guarding and adequate clearances.

The HSE advise a risk-based approach. Operators have to assess:

- the degree of the hazard, and
- the likelihood of access to the nip point.

In the former, the degree of hazard might be determined by the weight of the conveyor when stationary. If a person cannot lift the belt free of an idler using one hand, then control measures might be needed. In the latter of these, distance between persons and nip point is important.

A number of standards of guarding now exist around the world that set out detailed and specific measures to address conveyor nip point guarding. Australian Standard AS 1755-1986[9.3] states that:

> Guards shall be designed to prevent injury to persons and shall be provided at every dangerous part of a conveyor normally accessible to personnel.

This is interpreted as meaning that it should be physically impossible to access moving parts, through, over, under or round the guard. The measure of 'accessibility' means that, unless there is a vertical distance of 2.5 m separating persons from moving parts, a guard should be fitted.

In the United States, ANSI/ASME B20.1[9.4] provides safety guidelines for design, construction, installation, operation and maintenance of conveyors. These guidelines call for the application of guards at nip points.

For many years, BS 7300:1990[9.5] identified hazard points, and appropriate measures to deal with them, giving considerable detail with regard to guard design and where they should be fitted relative to the rotating part being guarded. A diagram of nip points included in this standard, perhaps unsurprisingly, has a striking resemblance to one found in AS 1755.

BS 7300 has now been superseded by BS EN 620:2002[9.6] as part of the programme of writing harmonised standards to cover all EU States. Section 5.1

of this standard, 'Measures for Protection against Mechanical Hazards', specifies similar detailed design features to those found in BS 7300.

It is clear from this brief review that accidents of this type continue to be commonplace, wherever in the world belt conveyors are used. Their frequency can only be combated by physical exclusion of personnel from hazardous areas, by programmes of training continually refreshed and by eternal vigilance on the part of those that supervise maintenance operations around belt conveyors.

9.3 Stored energy

The UK Health and Safety Executive (HSE) conducted a detailed survey of accidents associated with belt conveyors over the five year period between 1986 and 1991 and published their general findings in 1993.[9.7] One area that received special consideration was the effect of stored energy present in a belt when under tension. The major safety issue is that tension cannot always be seen by people working on a stationary belt, but to those involved with maintenance, cleaning spillage, and replacing or repairing joints, tension can prove dangerous. Key matters in these circumstances are the provision of adequate training of maintenance personnel and the provision of method statements. In every case, it stresses that the drive should be isolated before any work commences. The report also stresses the need for high capacity, properly designed pulling equipment clearly marked with a safe working load.

The report discusses the problem of a hold-fast situation in which very high tensions can be stored in a belt between drive and the affected area if the drive is not able to run back to release tension. Once the blockage is cleared, tensions either side of it in the trapped belt can equalise resulting in rapid, unexpected movement of the belt drawing personnel working on the blockage into a nip point situation.

People have been killed and badly injured when entering belt loops and other tensioning devices to maintain or clean them without ensuring that, once cleaned, the tension would not activate the device despite the belt not moving at the time. The HSE report called for means to be designed to allow safe removal of spillage from belt tensioning systems without the need for personnel to enter into hazardous areas.

9.4 Man-riding

Belt conveyors offer, potentially, a flexible and convenient method of transporting personnel but this activity can present a safety hazard. The question of dealing safely with personnel transport on belt conveyors can be dealt with in two ways. One way is to prevent it entirely, thus eliminating the hazard. This has been the approach in many industries and in many countries, most notably the coal mining industry in the United States where there is no culture of man-riding on conveyor belts and flexible and reliable alternative systems of rail and free-steered transport systems exist in mines. Man-riding is also not normally found on surface installations of conveyors in the UK.

Conversely, belt conveyors have been used for man-riding in the UK and throughout Europe and further afield in mines for many years. They offer the advantage of flexibility in selecting travel times, and are especially useful for the casual traveller such as belt inspection personnel or maintenance workers travelling mid-shift, not having to wait for a man-riding train. When suitably arranged, they also allow inbye travelling as well as travel outbye. One paper[9.8] on the subject of transportation in UK mines points out that little additional equipment is required to convert existing mineral conveyors up to man-riding standards. It further asserts that when two way (top and bottom belt) man-riding is installed, the improved maintenance standards required also ensure good coal clearance facilities.

The following narrative written by a UK coal miner was taken from the Internet and summarises well some of the factors associated with belt man-riding safety.

> Getting on the moving conveyor belt was an art in itself, if a person just stepped straight onto the belt they would be thrown backwards because the momentum of the belt took their feet away. The trick was to distribute your body weight correctly so that you were leaning forward as you stepped on the belt. It was then possible to adjust your position safely. Sometimes you had to throw yourself onto the belt and quickly lie flat due to height restrictions in certain roadways. The precise method depended on many things, the speed of the belt, the slope (on a drift belt), was there coal on the belt, or was the belt wet and slippery.
>
> Successfully negotiating the difficulties of boarding the belt did not bring to an end your problems, you then had to find a way of alighting the belt at the outbye end. A place had to be found with sufficient height and space, this was not always possible and you had to make do with what was there.
>
> Belt hangers were very handy, you just grabbed one and stepped off. But if it was low, you grabbed one and swung yourself off the belt as best you could. If you had any personal equipment with you, it was not always possible to get it off with you, so you had to run after it and retrieve it before it was lost over the conveyor head end.
>
> Armoured cables were also a godsend when it came to alighting the conveyor, these cables were suspended very close to the roof on cable hangers and stretched all the way up the roadway. You could always tell the most popular and probably the best place to alight by observing how dusty the cable was, if it was clear of dust then many others used that place and therefore you knew it was alright.
>
> Not everyone was always successful, occasionally, very occasionally, someone went over the head end and landed very uncomfortably on the next conveyor. Fortunately for them they were not seriously injured and they gave everyone else who heard about it a good laugh.
>
> (extracts from www.pitcraft.net/taxi)

Everyone associated with coal mining in the UK had to become adept at 'riding the belts'. In the mid-1960s one of the authors obtained work experience at a mine in the Cannock Chase coalfield. One of the first jobs was as assistant to the conveyor belt patrol man who travelled the extensive mine workings repairing joints and replacing idlers as necessary on the conveyor system. In order to travel the mine, conveyors were ridden, boarding and alighting at intermediate points where work needed to be completed. On one memorable occasion, it was necessary to get off the moving belt, where no hangers or cables were available to hold on to. To achieve this, the 'master' put both hands flat down on the belt and vaulted off, landing on his feet alongside the belt. The 'apprentice' followed suit, jumping just a little too high. He struck his back on the roof, fell back down on to the edge of the moving belt which deposited him in an undignified heap on the floor. No physical injury occurred, just embarrassment and much injured pride.

Several years later, the 'apprentice' had become a fully fledged management trainee. One day, he was riding along a gate road conveyor belt, alone, away from the coal face when his electric cap lamp cable became looped around some obstruction protruding from the side of the roadway. The cable broke, and suddenly he was in total and complete darkness but still travelling on the conveyor. This time belt hangers came to the rescue. He blindly held out his arm, grabbed the first hanger and stepped off on to the belt structure he hoped was there. Luckily, this was achieved successfully without injury, but all in complete blackness.

Much of the man-riding described by these contributions was 'illegal'. The best way to deal with it was to make it legal and provide facilities to both get on and off the belt with good clearances at well lit points. British Coal Corporation re-issued a set of Codes and Rules (CR/13) to govern the design of conveyors used as man-riders in 1994,[9.9] superseding and combining a number of earlier documents. It defined minimum clearances, maximum conveyor speeds, limiting gradients, conveyor construction standards, standards for boarding and alighting stations, notices, safety devices and inspection regimes. As the former British Coal Corporation no longer exists and its publications are unavailable, a précis of the guidance set out in this document is included in Appendix 3.

Even 'legal' man-riding could present problems for the casual user. An engineer friend of both authors broke his leg when boarding a legal man-riding conveyor awkwardly. Despite accidents like this, man-riding has been a very successful and safe method of transporting people large distances underground when alternative systems were not available.

The increased speed of many modern conveyor installations has also mitigated against widespread man-riding on conveyors that also carry mineral. The authors have become aware that specific man-riding belt installations can be found in Australian and South African mines, there is a South African standard that relates specifically to man-riding on belt conveyors,[9.10] but man-riding on mineral-carrying conveyors is not thought to be normal practice in these countries.

The Mines Inspectorate in the UK has always taken a deep interest in man-riding on belt conveyors. HSE Books published a 'Topic Report' on the safe use of belt conveyors in 1993, that covered, amongst other subjects, man-riding.[9.7] It published gruesome stories of accidents in coal mines that had resulted from illegal use of belt conveyors for travelling. In one incident, two men travelling illegally had passed through an airlock on a belt. The first man became stuck in the small gap allowed for the belt, the second ran in to him, and both were engulfed by coal on the belt that asphyxiated them! In another horrible accident, a surface worker rode a feed belt for a crusher that took him under the safety trip-out wires and, fatally, through the crusher!

In some respects, the HSE Topic report also ran closely alongside the British Coal's document.[9.9] It listed the potential hazards to be aware of when operating a man-riding conveyor. It mentioned the need for discipline whilst travelling on a conveyor, and pointed out the hazards associated with illegal man-riding. The HSE authors were clearly concerned with man-riding at the same time as mineral transport. They stressed the need to ensure that the bed of mineral does not result in unacceptable clearances, that travelling on mineral can be done safely (a slight contradiction here with the BCC Codes and Rules), and a concern that mineral does not roll along conveyors on gradients or create dusty conditions or blockages. They also stressed the need for adequate control of belt speed and braking when riding on inclines. It recommended that belt widths should not be less than 1.05 m (750 mm in CR/13), but its observations on maximum speed are in keeping with those of CR/13.

Translations of codes of practice from mines in France and Germany have also been examined by the authors.[9.11,9.12] These documents demonstrate that man-riding by belt conveyor was a normal mode of personnel transport in mines in both those countries. The Codes expressed similar restrictions on belt speed and clearances as those found in UK documentation (see Appendix 3).

The increased application of lump-breakers, coal sizers, and bunkers underground clearly had a large impact on the application of conveyor systems feeding into them for man-riding purposes. British Coal had produced guidance on the use of these features both at the surface and underground in 1990.[9.13]

The CR/13 document stated that no conveyor that delivers into a bunker or crusher should be used for man-riding. The Topic report[9.7] recognised the incompatibility of conveyor man-riding and crushers. However, the earlier British Coal lump-breaker guidance stated that 'the siting of lump-breakers after man-riding belts should be avoided wherever possible'. To prevent legal man-riding leading to accidents, alighting stations were to be sited a minimum of 100 m before the breaker installation. Furthermore a pull cable that shut down both conveyor and lump-breaker was to be installed directly over the conveyor along that 100 m distance.

Means had to be found to prevent even illegal man-riding or persons inadvertently drawn onto a feed conveyor from entering a bunker or lump-breaker. This was achieved by fencing or by excluding workers from sensitive areas as much as possible. The guidance recommended that the guards fitted over in-feeds should exclude persons but should also be designed to be unsuitable as working platforms or as places to store materials and equipment. Finally it recommended that adequate, clear warning notices should surround the site.

9.5 Materials handling by belt conveyor

In the restricted world of the underground mine, transport does not only include the movement of minerals in one direction to the surface and personnel in both directions through the mine workings, but also materials and equipment in to the working places.

Once again the belt conveyor has a part to play in this process, especially in the final few hundred metres from the rail head or end of haulage to the new working places. Mining by definition involves constant geographical movement of the focus of working activities. The belt conveyor is quickly installed or extended to follow this progress.

By its very nature, the bottom belt moves inbye as mineral is removed on the top belt outbye. Therefore, it has been common practice to use this valuable resource to move smaller items of consumable materials used in mining in along the bottom belt. This can be achieved safely and without incident if installation standards are high and personnel are familiar with the activity. The HSE Topic Report[9.7] recognises that this activity is part of normal mining life and reminds the reader of the statutory requirement for a mine manager in the UK to make rules for transport activities under the Mines and Quarries Act 1954. These are known as the 'Manager's Transport Rules'. For transporting materials on belts, the Topic report recommends that the rules should require the following:

- suitable clearances at loading and unloading stations
- conveyors to be stationary during loading and unloading
- ploughs to be fitted at bottom belt unloading stations (to prevent materials over-running into the return pulley nip point)
- materials to be positioned centrally within the width of the belt
- all materials loaded are accounted for at the unloading point
- pedestrians should be excluded from travelling with the load or alongside the belt during transport operations.

9.6 General comments

Chapters 8 and 9 have looked at the main safety considerations associated with operating belt conveyors. The authors have noted how the interaction between people and the belt conveyor can lead to hazardous situations. Furthermore, there is nothing particularly new about these types of incident. Examples have been quoted from around the world covering a period dating back to 1950. It is rather sad to note that the same accident types happen time and again, whether they be fires associated with belt conveyors in underground mines and tunnels or the inadvertent drawing of persons into nip points when they feel it necessary to overcome the protection systems put in place to exclude them from the danger area.

It should be too pessimistic a view to state that these events will always happen. However, until operators of belt conveyors truly engage with the fact that a belt conveyor can represent a real hazard, and that there are risks associated with mixing people, especially untrained or unskilled ones, with belt conveyors in often remote locations, then it is hard not to see a whole new group of accidents statistics and case studies being available for writers of books such as this in years to come.

9.7 References

9.1 Hollyoak M 'Why are nip guards required on belt conveyors in mines but not in quarries?' Seminar on Bulk Handling Conveyors, Midland Institute of Mining Engineers, Sheffield May 2006.
9.2 Conveyor Equipment Manufacturers' Association (CEMA) *Belt Conveyors for Materials Handling* 6th edition 2005.
9.3 AS 1755–2000 'Conveyors – Safety requirements' Standards Australia.
9.4 ANSI/ASME B20.1 'Safety standards for conveyors and related equipment.' American National Standards Institute.
9.5 BS 7300:1990 'Code of Practice for Safeguarding of hazard points on troughed belt conveyors' British Standards Institution.
9.6 BS EN 620:2002 'Continuous Handling Equipment and Systems – Safety and EMC requirements for fixed belt conveyors for bulk materials' British Standards Institution.
9.7 Health and Safety Executive *Safe Use of Belt Conveyors in Mines – Topic Report* HSE Books 1993.
9.8 Shepherd C and Gilbert S 'A Review of underground transport systems in the Barnsley area of the NCB' *Mining Engineer* 225, June 1980.
9.9 British Coal Corporation Codes and Rules CR/13 – Underground Belt Conveyors'.
9.10 SANS 10266:1995 'The safe use, operation and inspection of man-riding belt conveyors in mines' Standards South Africa.
9.11 National Coal Board Translation M24509 of German document Code of Practice of man-riding on conveyors *Betriebsempfehlungen* No 17, SKBV, August 1978.
9.12 National Coal Board Mining Research and Development Establishment Translation 1055 of French document Technical publication No 2, HBNPC, 1979.
9.13 British Coal Corporation Technical Notes of guidance for the installation, operation and maintenance of lump-breakers underground and at the surface BCC March 1990.

10
Maintenance and monitoring

10.1 Introduction

In Chapter 2 we discussed the economics of conveyors as against other methods of moving bulk materials. In Chapters 8 and 9 we have highlighted the safety issues associated with belt conveyor systems. Installing a conveyor can be a major investment, and this investment deserves appropriate care and maintenance if it is to work properly and to give long, efficient and safe service. Often the largest item of expenditure involved in an installation is the conveyor belt itself. It is therefore of primary importance that the belt in particular is looked after properly at all stages in its life. This chapter is concerned with how a conveyor is maintained and how it can be monitored to maximise life and efficiency. We have tended to concentrate on how the belt itself should be managed.

10.2 Supply, storage and handling of belts

Conveyor belts for bulk materials handling are generally supplied as single- or double-coiled rolls. In most circumstances wood or steel cores, having a central hole to accommodate a mounting bar, are inserted to assist handling of the rolls. The size of the core depends on the size of the roll, but ISO 5285 'Conveyor belts – Guidelines for storage and handling'[10.1] recommends cores be sized so as to accept 50, 100, 150 and 200 mm square mounting bars. The length of belting required and the size and weight of roll that can be accommodated are specified by the purchaser, and will be dependent on the handling and installation requirements and limitations at the installation site. Naturally, the length of belt in a roll depends on the size of the roll, the thickness of the belting and the core size. Typically, belt may be supplied in lengths up to 400 metres, roll diameters up to three metres and roll weights up to ten tonnes. The length of belt per roll needs to be such as to minimise the number of joints needed in the installation.

Although conveyor belting may appear to be robust, it is, in certain circumstances, quite easily damaged. The principal agents of damage are attack by ultraviolet (UV) light, ozone, heat, oils or chemicals, and abuse by human beings. Covers can be formulated to have special properties such as heat, ozone or oil resistance (see Chapter 5) but the standard grades of rubber used for covers are not resistant to attack. UV light, ozone, oils, chemicals and heat can all cause degradation of rubber compounds. Conditions of storage need to be such that the belting is protected from these agencies. UV and ozone attack appears as cracking and crazing of the rubber, and belts should be stored away from direct sunlight or sources of ozone such as electric motors. Rubber may be attacked by oils or solvents and will readily swell and degrade in contact with them. Heat also degrades rubber and reduces life expectancy of the belt. Ideal storage temperatures are between 10 and 25 °C. Belts also need to be stored in conditions free from damp or high humidity to avoid the possibility of mildew attack. Some types of belt become much stiffer if stored below 0 °C and may need to be allowed to warm up before being handled.

Belts may easily be damaged by lifting equipment. Care needs to be taken to avoid direct contact of lifting equipment, such as chains, with the surface or edges of a belt. In addition the carcase may be damaged internally by folding or bending the belt over a small radius. A common cause of damage during installation is dragging the belt over sharp edges.

ISO 5285 gives guidance on conditions for storage of belts to avoid damage and on appropriate lifting procedures, handling with fork-lift trucks and hints on installation procedures.

10.3 Belt tracking or training

If a conveyor is to give satisfactory performance, it is extremely important that the belt should run true to its intended course, i.e., that it should track or train properly, so that the belt does not become damaged and spillages of the material carried do not occur. We have mentioned in Chapter 3 the fact that in recent years it has become more common for conveyors to contain (intentionally!) horizontal curves, but by far the best way to run a conveyor is still in a straight line. For this to happen it almost goes without saying that the structure supporting the idlers, belt and drives must be straight and true. All of the pulleys and idlers must be at right-angles to the centre line of the conveyor and centred on the conveyor centre line.

If pulleys or idlers are not perpendicular to the centre line of the conveyor, the belt will move towards the side of the pulley or idler that first touches the belt. The idler mounting brackets must be properly horizontal so that single return idlers and the centre idler of troughed idler sets are horizontal. If these idlers are not horizontal then the belt will move towards the higher side. Extreme care is needed when making adjustments to the position of idlers or pulleys if the belt is running. Chapter 9 has warned of the speed at which clothing or limbs can be pulled into nip points. The belt must be installed in the centre of head and tail pulleys, and should trough properly when empty, so that it touches the centre idler on the troughed side, since it is this idler that steers the belt. Take-up devices should be free to move centrally on the centre line of the conveyor and pulleys in the take up must remain at right-angles to the centre line of the conveyor.

On a new installation, or if a new belt has been installed on an existing installation, it is advisable[10.2] to run the belt loaded as soon as it has been installed, then stop it overnight still loaded. This is said to speed the running-in process and the stabilisation of any unequal stresses across the width of the belt arising from troughing. Tracking should start on the return side close behind the head end or the take-up and move towards the tail end. Tracking of the carry side should start at the tail end and move towards the head.

As highlighted in Chapter 4, off-centre loading of the belt as the material being transported is deposited on the belt at the loading point can cause a belt to run off track, as can build up of material on the idlers or on pulleys. Edge wear leading to the belt edge stretching, or ingress of water leading to the belt edge shrinking can also cause the belt to run off line.

If the belt runs off track at one particular point on the belt then either the belt or a belt joint is not straight. However, if the belt runs off at one point on the structure then the structure or idlers are at fault. Adjustment of idlers to bring the belt back into line in the latter case should be done some distance before the position where the belt runs off. Reference 10.2 advises that this distance should be between 1.5 and 6 metres. Tilting the wing idlers of troughed sets or vee return idlers forward slightly is a recognised method of assisting the alignment of a belt,[10.3] although as we have seen in Chapter 3, this does increase the power required to drive the conveyor. As we have remarked in Chapter 4, the use of crowned pulleys can assist in aligning the belt, but will increase stresses in the belt.

In underground mines movement of the roof walls and floor of roadways or tunnels often occurs and can cause the structure of a conveyor to move out of line, and regular inspections and adjustments to conveyors are required. In any installation, whether underground or otherwise, periodic inspections to check for loss of alignment of the conveyor should form part of the maintenance schedule.

When all adjustments have been exhausted and the belt still runs off track, it may be necessary to install self-aligning idlers that work on a pivot and that will bring the belt back into line, or to install guide rollers that contact the edges of the belt and bring it back to its intended course.

10.4 Optimising belt life

While proper attention to tracking will help to optimise belt life, it is generally true that belt life is optimised if the stresses imposed on it are minimised. These stresses may arise from the design of the installation, or from its maintenance. Many of these sources of stress have already been mentioned in earlier parts of this book. However, it is worth reiterating some of them.

While the factor of safety (service factor) governs the overall tensile stress imposed on the belt, the design of transitions, convex and concave curves (Chapter 3) and of horizontal curves (where these are intentionally included) can impose edge or centre stresses that will shorten the belt life. Troughing of the belt creates longitudinal creasing of the belt, transverse flexing and non-uniform longitudinal stresses. All of these are lower for lower troughing angles and the choice of the lowest possible angle of trough consistent with the design capacity of the conveyor will maximise belt life. The possibility of the belt being pinched between the horizontal and wing idlers has been mentioned in Chapter 4, which notes the potential advantage of staggered idler sets in this context. BS 8438[10.3] advises that the maximum gap between the centre and wing idlers should be 10 mm.

The design of loading points has been discussed in Chapter 4, which notes that the material needs to be fed onto the belt in a controlled manner and at a speed compatible with the speed of the receiving belt to minimise impact and cutting and gouging of the belt covers. Similarly Chapter 4 discusses the design of belt cleaners, which, if not positioned and maintained correctly can cause belt damage. Ineffective belt cleaners can also allow material to build up on pulleys or idlers and this may cause belt damage or mis-tracking.

Severe mis-tracking of the belt can result in contact of the belt with structure or other objects and damage to the belt edge. This damage can be compounded if strands of fabric from the belt become wrapped round idlers, with the potential for jamming them, increasing drag or causing heating. BS 8438 gives guidance on clearances that should be maintained. These clearances have set values of 50, 75 and 100 mm for belts of 400–650 mm, 750–1400 mm and 1600–2000 mm in width respectively. Tramp metal can cause major problems. In underground coal mines roof bolts, steel rods used to reinforce and stabilise the strata, have been known to become jammed in chutes and to cause extensive ripping of belts, particularly steel cord belts, which are prone to this type of damage.

The maintenance of idlers is of great importance. Idlers that are dirty or are not rotating freely can cause the belt to run off track. Idler bearing failure, leading to an idler jamming, is not only a potential source of belt wear as the belt rubs over the stationary roller but also leads to increased power requirements, and is potentially a source of fire.

We have discussed in Chapter 8 the potential for causing fire of a stalled belt rubbing on a drive pulley that continues to revolve. It is worth pointing out that various types of slip detector are available that can be used to shut down a conveyor if relative slip exceeds a given figure.

10.5 Monitoring the condition of belts

In addition to visual inspection of belts, which can reveal a great deal about belt and joint condition, various destructive and non-destructive tests are available that can both reveal faults that are not detectable by visual examination and quantify belt and joint condition.

It was fairly common practice in the UK coal mining industry during the 1980s and 1990s to take sample lengths of belting that included mechanical or spliced joints from critical conveyors and send these to laboratories for examination and testing. The testing could reveal the extent of cover wear and carcase damage and quantify condition through tensile tests on both the belt and the joints following the methods outlined in Chapter 7, so that the remaining factor of safety could be determined. However, non-destructive methods of testing, such as are described by Harrison,[10.4] were in use on some of the major steel cord belt installations. Harrison discusses the development of testing methods that could be applied to moving belts *in situ*. He mentions several technologies that became available to examine belt condition, including magnetic, eddy current, electromagnetic X-ray and vibration sensors. He goes on to describe a system that had been successfully developed and proven, which had the objective of detecting broken cords or cords damaged by corrosion through the ingress of water through cover damage. This system also had the capability of detecting and quantifying the condition of splices following the characterisation of the particular splice geometries initially. From the results of these belt examinations it was possible to compute a revised factor of safety for the belt based on the local damage found.

X-ray examination of the splices on the Gascoigne Wood steel cord belt installation were used by one of the authors and his colleagues of the time, following the incident reported in Chapter 6 Section 6.3.2 in which a spliced joint failed because of incorrect curing. The tests were carried out on the splices over two weekends in the drive house at the surface in the area of the splicing station that was built into the installation close to the drive. No other problem splices were detected, but the exercise proved the value of the technology as far as British Coal were concerned.

Non-destructive methods of examining the integrity of fabric belts have been developed, but the authors are not aware that they were used in the UK mining industry, and as Harrison remarks, they were not considered to be cost effective.

As we have said previously, steel cord belts in particular can be prone to ripping if a sharp object gets pushed into the belt. In Chapter 5 we mentioned the inclusion of wefts or breakers in steel cord belts to protect against rips. Other systems of rip detection and protection have been devised and are in common use. The systems work by the embedding in the belt of sensor loops that pass detectors as the belt runs. If a loop is broken in the event of a rip then the detector picks this up and the belt can be stopped.

On critical conveyor installations general machinery monitoring procedures such as are followed in other industrial sectors are often installed. These include vibration monitoring of pulley bearings and the use of wear debris analysis of oil in gear transmissions. Infra-red detectors and thermal imaging cameras have been used on conveyors in coal mines to pinpoint hot bearings in idlers or pulleys.

10.6 References

10.1 ISO 5285:2004 'Conveyor belts – Guidelines for storage and handling'.
10.2 *Fenner-Dunlop Technical Manual* Fenner-Dunlop Conveyor Belting Europe Limited, Hull, UK.
10.3 BS 8438:2004 'Troughed belt conveyors – Specification'.
10.4 Harrison A '15 years of conveyor belt monitoring evaluation' *Bulk Materials Handling* Vol 16, No 1 1996.

11
Case histories

> The belt conveyor being installed (represents) a 1000% increase in the power of the drive ... a 500% increase in the strength of the conveyor belt and the belt speed is double that of the majority of belt conveyors.
>
> H. D. Hoy and P. A. Lowry, *Mining Technology*

This book has been written to assist mine or quarry operators who are considering installing belt conveyor systems and who require basic engineering information to assist with their decision-making process. It has also been written for students in those engineering disciplines that will ultimately lead to employment in the mining and quarrying industry. Much of the information given is generic in nature. However, this final chapter contains a number of case histories that illustrate various points made elsewhere in this book, and also in its final section, what we describe as 'The biggest and the best', a number of illustrations that demonstrate the major role that the belt conveyor can play in the minerals extraction world today.

To begin the chapter, we have chosen to describe two examples of conveyor systems that were designed and installed almost thirty years ago in the coal mining industries of the UK and Germany, that were trail-blazers in their time and which remain significant engineering feats to the present day. Details of a third coal-mining conveyor system that commenced operations in 2007 are also included to demonstrate the tremendous advantages that a belt conveyor can bring to a surface mining operation in an environmentally sensitive area.

11.1 Selby mine

The first of the Selby Coalfield mines in North Yorkshire, UK was planned during the late 1970s and commenced production in 1983. The coalfield was designed to a plan that required that five individual, but physically connected, deep mines would load all of the coalfield output onto conveyors placed in two parallel 'spine roads' (Fig. 11.1). As the name implies, these two tunnels ran down from the surface at Gascoigne Wood at a gradient of 1 in 4 (25%) to a grade

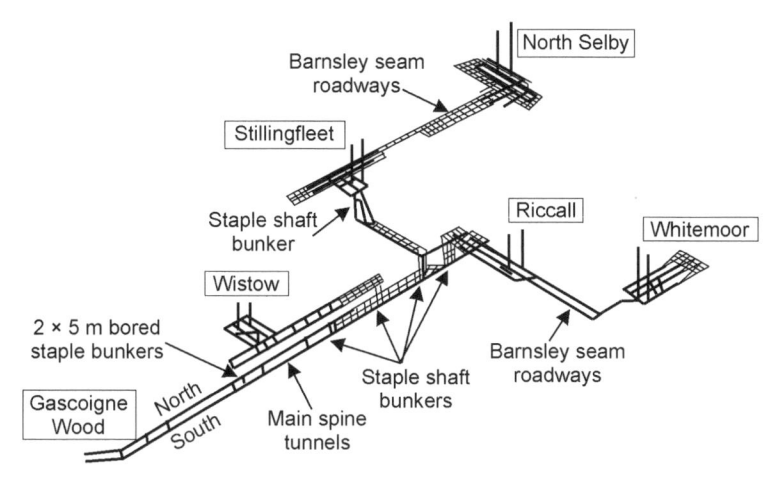

11.1 Selby complex schematic diagram (former NCB photograph – published by kind permission of Department for Business, Enterprise and Regulatory Reform).

approximately 70 m below the Barnsley seam and then bisected the underground workings from south-west to north-east forming the axis of the coalfield.

The original planned length of these spine roads was 14.8 km but in the event, following changes to the plans as development progressed, they reached a distance of 12.2 km. Eleven strata bunkers were to be built, each some 7.315 m in diameter to store each mine's output and allow a controlled feed onto the conveyors in the spine road. The duty for the conveyors installed in the spine roads was simple, yet challenging.

> ... to provide in each of the spine roads a conveying system capable of handling the full output. In simple terms this is 2000 tonne/h over a distance of 15 000 metres and with a vertical lift of 1000 metres.[11.1]

As the conveying duties were at the edge of existing technology at the time, the mine operator decided to use different types of conveyor for each spine road.

A cable belt design was installed in the North Tunnel. This eventually extended to a distance of some 9.5 km and was primarily used to convey the output from Wistow mine, the first mine in production. Details of the operation of the cable belt principle can be found in Chapter 5. The belt was powered by 2 × 4100 kW motors through 6 m diameter Koepe wheels that drove the 59 mm diameter wire ropes on which the 1 m wide flat belt rode. This belt was originally designed to convey 2000 tonne/h at full speed, but it was reported[11.2] that vibration of the ropes caused spillage in the roadway, coupled with a relatively high incidence of pulley failure. Additionally, intermediate loading from bunkers onto the conveyor belt had also caused spillage problems. As a result, a lower capacity of 1500 tonne/h was adopted whenever possible in practice.

When the final 2.5 km of the tunnel was driven and an extension of this conveyor was required in the early 1990s, the mine chose to install a separate

160   Belt conveying of minerals

tandem conveyor loading onto the cable belt.[11.2] The argument was given that to extend the cable belt would be to reduce its capacity. A tandem conveyor would also give a better, centralised feed when delivering onto the cable belt. Less tunnel drivage would be required than if the cable belt were to be extended, because the necessity for rope tension and storage arrangements would have been avoided.

The tandem conveyor was 1350 mm wide, utilising 3150 N/mm (Type 18) belting. This low stretch belt had a 5 mm thick upper cover and a 4 mm thick lower cover. This was expected to give a 15-year operating life. This was one of only two occasions when this type of belt was applied in UK coal mines[11.3] (see Chapter 3).

The South Spine conveyor was perhaps the groundbreaker at the time (Figs 11.2 and 11.3). It was a steel cord conveyor, 1300 mm wide, driven by 2 × 5050 kW dc motors directly coupled to a driving pulley. The drive was sited at the surface at Gascoigne Wood. The conveying speed was variable up to a maximum of 8.4 m/s. The steel cord belt was 28.3 mm thick with 57 cords each 13.1 mm in diameter at 22.25 mm pitch. The belt was reported to have a factor of safety of 5:1 and was graded as ST7100.[11.4] The coal from the mine bunkers was fed onto the main conveyor by way of an accelerator conveyor that ran at up to 75% of the variable speed of the main conveyor.

The duty was challenging. As one observer commented at the time of the installation of the conveying system.[11.3]

> The belt conveyor being installed (represents) a 1000% increase in the power of the drive, to any that are operating within NCB mines. There is a 500% increase in the strength of the conveyor belt and the belt speed is double that of the majority of belt conveyors.

The South Spine conveyor operated throughout the life of the Selby coalfield through to its closure in 2004, but not without some problems.[11.5] In August

11.2 Selby South Spine conveyor (former NCB photograph – published by kind permission of Department for Business, Enterprise and Regulatory Reform).

Case histories 161

11.3 Selby South Spine conveyor drive (former NCB photograph – published by kind permission of Department for Business, Enterprise and Regulatory Reform).

1987, vibrations were noted from the conveyor drive bearings and the conveyor had to be stopped. The drive was stripped down to examine the bearings. This was a major task as the drive shaft and pulley were all one forging weighing some 90 tonnes, which increased to 125 tonnes when bearings and clutches were included. It took 36 hours to get into a position to remove the bearing from the shaft. Each bearing was 1.5 m in diameter and contained 42 rollers, each weighing 9.5 kg and measuring 140 mm in length with a diameter of 105 mm. It was found that one of the rollers had fractured, leading to bearing race damage. There was a second similar occurrence in February 1988. On investigation it was found that another roller had fractured. This was thought to be a quality problem and various changes to manufacturing processes and steel specification were put into place. Because, at the time, the output from the Selby mines was a considerable proportion of overall British Coal Corporation output, the financial fallout from this incident was considerable. The Corporation's estimated losses were in the region of £1 million a day. The importance of reliability of the conveying system to the overall financial success of the mining operation had been amply demonstrated.

For the record, the Selby complex reached its planned annual output of 12 m tonnes per annum production in only one year, but the conveying systems remained a significant feat of engineering throughout.

11.2 Prosper-Haniel

Another significant example of the application of high-capacity conveyor technology was being installed across the North Sea in Germany. Ruhr district coal mines were, in general, working reserves of coal well to the north of the original access points to the coalfield. This had led to convoluted transport systems developing as mining moved away from the shafts, over distances of many kilometres and at varying levels back from the working areas to the surface preparation facilities. In 1986 the Prosper-Haniel mine of RAG in Germany was re-organised and a decision was made, unusually for German mines, to drive a surface drift to convey coal from the working levels to the surface.[11.6] The conveying system installed had a length of 3743 m and lifted mine output 783 m at a gradient of 13 degrees (23%) to the surface at a rate of 1800 tonne/h. An additional duty for this same conveyor was to transport washery discard back into the mine for disposal at the rate of 1000 tonne/h. As with Selby, the belt was steel cord in construction with the main drive at the surface.

The power requirement for this system was 5880 kW to transport the coal only, at a variable speed up to 5.5 m/s. Interestingly, two-thirds of this power was needed to raise the payload, the other third being employed to overcome friction. The drive was fitted with 2×3100 kW rated motors onto a single pulley, 2.2 m in diameter. The drive arrangement was chosen such that one motor alone could lift the coal output if necessary, as long as washery discard was being transported into the mine at the same time to assist with the effective load.

The steel cord belt used was rated at ST7500. The supplier of the belting, Phoenix Conveyor Belts of Hamburg claim that this is the strongest underground conveyor belt in the world. It was 1400 mm wide and 34.5 mm thick, with a carcase composed of 72 steel cords of 12.5 mm in diameter set at 19 mm centres. The cords were covered with textile cross-armouring (a breaker) to reduce the risk of horizontal tears. A static factor of safety of 6.14 was achieved. We have been told that this installation has now been working at Prosper-Haniel for over 20 years successfully performing the duty for which it was installed.

11.3 ATH Resources

ATH Resources are a surface coal mining company that operates a number of sites around the town of New Cumnock in Ayrshire, Scotland. Although deep coal mining was a feature of this district for much of the 19th and 20th centuries, deep mining has now ceased and only the scars of these operations, in the form of occasional waste dumps, can be seen in the surrounding area. What remains is a beautiful and quite hilly moorland landscape that contains a major salmon river, the River Nith and areas of forestry, and that possesses the potential for increasing tourism. However, significant coal reserves remain quite close to the surface and ATH are working these.

Surface coal mining in the UK is constrained severely by the need to satisfy environmental requirements. One of these key constraints is to design a satisfactory arrangement to transport the mined product away from the mines. The rural area around New Cumnock is not blessed with good quality roads, but there is a railway system and a rail loading facility. The problem facing ATH has been to transport coal from their sites to the rail loading facility. This has been achieved by building what, at 12.2 km in length, is claimed to be the longest belt conveyor system in Europe.

This has not been a straightforward installation due to:

- the exposed and hilly terrain
- the ground conditions under some of the route
- the remoteness of the route from existing services
- the need to cross roads, a river and a 'Special Protection Area'
- the need to provide electrical power all along the conveyor system route
- the difficulties associated with access to the route during the construction of the system and
- the requirement for the conveyor to blend into the surrounding landscape.

Operationally, this is not an exceptionally high capacity conveyor by some international standards at 500 tonne/h, but it does include complex curves with both horizontal and vertical radii. It is also exposed to the wet and cool climate of south-west Scotland.

ATH operates several sites in this area and decided to transport the coal mined at the site where the starting point of the conveyor was to be situated via two other sites that were situated at intermediate points on the route to the rail loading site. This produced a route that required the conveyor system to negotiate several horizontal curves and several sharp changes in direction.

The conveyor system consists of ten separate individual conveyors, the longest of which is 2.32 km in length. Several of the conveyors incorporate horizontal curves, the tightest of which has a radius of 600 m and the largest 1000 m. Three of the conveyors lift, i.e., the head is higher than the tail, while the remainder fall (the tail is higher than the head). The individual conveyors are of relatively straightforward design, using single pulley drives and gravity tower tensioning arrangements. As will be seen from Fig. 11.4, the line stands are mounted on wooden sleepers, which in turn sit on ballast. This provides stability for the structure to preserve alignment. The whole length of the conveyor, including the transfer points, is encased in green coloured cladding to reduce the visual impact of the system as well as to provide protection and containment of dust. Public access to some parts of the conveyor system run necessitates particularly high standards of guarding and containment. These can be seen in Fig. 11.5, which shows one of the conveyor drives. Figure 11.6 shows the conveyor mounted on a gantry to enable it to cross a valley in a Special Protection Area. The belts are steel cord, ST1250 which are 900 mm wide and run at 2.73 m/s. Tensions have been kept low to enable the conveyors to negotiate the horizontal curves. The final conveyor in the system that feeds the rail loading point incorporates a travelling tripper-stacker that produces a 165 m long stockpile (Fig. 11.7). The conveyor system is provided with a Supervisory Control and Data Acquisition (SCADA) system which enables ATH to access all

164   Belt conveying of minerals

11.4 General view of the ATH conveyor showing the line stands and sleepers (by courtesy of ATH Resources Ltd).

diagnostic information. Control of the conveyor system is from a central point that has an electronic data network to communicate with individual conveyors. The conveyor system has been estimated to remove around two million truck miles from the roads around the area and reduce carbon dioxide emissions by 3500 tonnes annually.

11.5 The ATH conveyor showing the guarding and one of the drives (by courtesy of ATH Resources Ltd).

11.6 ATH conveyor crossing part of a Special Protection Area (by courtesy of ATH Resources Ltd).

11.7 Delivery stockpile tripper-stacker at the railhead (by courtesy of ATH Resources Ltd).

11.4 'The biggest and the best'

Technology never stands still in any sphere of engineering and this holds true for belt conveying. Therefore any reference to a system being the longest, or having the highest capacity will inevitably become out of date quite quickly as another system comes into use somewhere in the world. The final part in this chapter therefore is written with some trepidation as the examples chosen will instantly age the book as they becomes out of date.

Phoenix Conveyor Belts has a website in which they list a number of conveyor installations using their equipment that they claim to be 'the biggest and the best'. Whilst the authors understand that other companies may take issue with these claims, some of the installations chosen and briefly described below, clearly represent the present high capability of the belt conveyor and demonstrate the technology that is now available to mine managers who are increasingly faced with the challenge of operating in geographically remote and inaccessible places in this world and transporting their product to a distant processing facility and thus the marketplace. However, the authors have not relied entirely upon the material published by Phoenix. The technical press also reports major new installations at regular intervals, a number of which are included in the brief descriptions below.

The longest single stage conveyor running in the world is generally thought to be the installation that takes between 800 and 1000 tonnes per hour of limestone and shale from the Lafarge Quarry in the East Khasi hills in the Indian province of Meghalaya to a cement plant at Chhatak, Sunamganj in Bangladesh, a distance of 16.5 km. The belt is rated ST2500 and is 800 mm wide. The longest conveyor system is claimed to be the 100 km long Fosbucraa installation running from the phosphate mines of the Western Sahara at Bu Craa to the coast.

Other systems claim some attention as the limits of conveying technology are stretched ever further.[11.7] The El Abra copper mine is situated in the Chilean Andes. The material from the primary crusher is transported from an elevation of 4000 m above sea level via a three-flight conveyor system, a distance of 15 km to a stockpile at an elevation of 3400 m. The longest flight of this conveyor system is 9.5 km long and has a drop of 510 m. This conveyor also includes a horizontal curve in its route. When running at full load, it has a capacity of 8140 tonne/hr at a speed of 6.1 m/s. The belt is rated at ST6800.

Los Pelambres Copper mine is located at an altitude of 3200 m, also in the Chilean Andes.[11.8] In order to transport mined ore from mine to concentrator, a belt system has been installed with a specification that takes belt conveyor engineering near to its limit. The capacity of the system is 8700 tonne/h at a speed of 6 m/s, the belt is ST7800, 1800 mm wide and 42.5 mm thick is claimed to be the world's strongest by its supplier, Phoenix. The conveyed distance includes a vertical drop of some 1400 m to the concentrator. The system includes three flights of 6 km, 5.3 km and 1.4 km in length. The drive pulleys for this system were also classed as the largest built to date at the time of installation (1999) at 2500 mm in diameter.

Batu Hijau gold mine is located in south-west Sumatra, Indonesia.[11.9] The overland conveyor from this mine that leads from mine site to concentrator has a length of 5.6 km with an overall drop from tail to delivery of 86 m. However, during its run there are various undulations including a 160 m downhill section followed by a 75 m rise. There are five concave and five convex vertical curves as well as a single 9000 m radius horizontal curve. The capacity of the belt is 8433 tonne/h at a speed of 4.73 m/s, the belt width is 1800 mm and the rating is ST4600.

These few examples are selected from many more whose details can be found in the technical press and on the world wide web. They serve to demonstrate that belt conveyor systems continue to grow in capacity and versatility, that they can be installed in hostile environments and that they can be built to navigate both horizontal and vertical curves and operate over long distances.

11.5 References

11.1 Massey C T and Dunn J 'Steel Cord Conveyor' *Mining Engineer* 274, July 1984.
11.2 P M Davies Final Conveyor Link in the Selby Complex *Mining Engineer* 368, May 1992.
11.3 Hoy H D and Lowry P A 'Gascoigne Wood – Eagle or Albatross' *Mining Technology* 77 891, November 1995.
11.4 R G Siddall 'The Selby Coalfield' *Mining Engineer* 326, November 1988 p. 221.
11.5 R G Siddall 'Selby – an update on progress' *Mining Engineer* 339, December 1989 p. 228.
11.6 Ketteler H 'Improving the infrastructure at Prosper-Haniel by means of a 3653 m long belt incline and a single surface preparation plant for centralised coal conveying and washing.' British Coal Corporation Headquarters Technical Department translation 0391.
11.7 Kahrger R 'El Abra – Two years later' *Bulk Materials Handling by Conveyor Belt II*, SME 1998.
11.8 Brewka C 'The Los Pelambres overland conveyor' *Bulk Materials Handling by Conveyor Belt III*, SME 2000.
11.9 Chan P K 'Design of the High capacity, Horizontally Curved Overland Conveyor at Batu Hijau' *Bulk Materials Handling by Conveyor Belt III*, SME 2000.

Appendix 1
Derivation of belt capacity

The belt capacity is derived by simple geometry from a diagram such as Fig. A1.1, in which there are three rollers of equal length and where

- α is the angle of surcharge
- β is the troughing angle
- W is the belt width (m), and
- Y is half of the length of the centre roller.

All calculation methods assume that the belt is filled uniformly along its length and that the load extends to within a distance x of the edge of the belt. This distance has been derived empirically and is expressed as a fraction of the belt width plus a constant, e.g.

$$x = 0.05.W + 0.025 \qquad \text{A1.1}$$

where all units are metres. The cross-section of the load consists of a trapezium ABCD and a segment of a circle ADE. To calculate the area of the trapezium we need to know the length Z along the wing idler to the limit of the load.

$$Z = \frac{W}{2} - Y - x \qquad \text{A1.2}$$

The area of the trapezium $Area.1$ is

$$Area.1 = 2.Y.\sin\beta + Z.\sin\beta.Z.\cos\beta$$

which reduces to

$$Area.1 = Z.\sin\beta(2.Y + Z.\cos\beta) \qquad \text{A1.3}$$

To obtain the area of the segment we need to know the radius R of the arc AED, which is given by

$$R = \frac{Y + Z.\cos\beta}{\sin\alpha} \qquad \text{A1.4}$$

The area of the sector OAED is

Appendix 1: Derivation of belt capacity 169

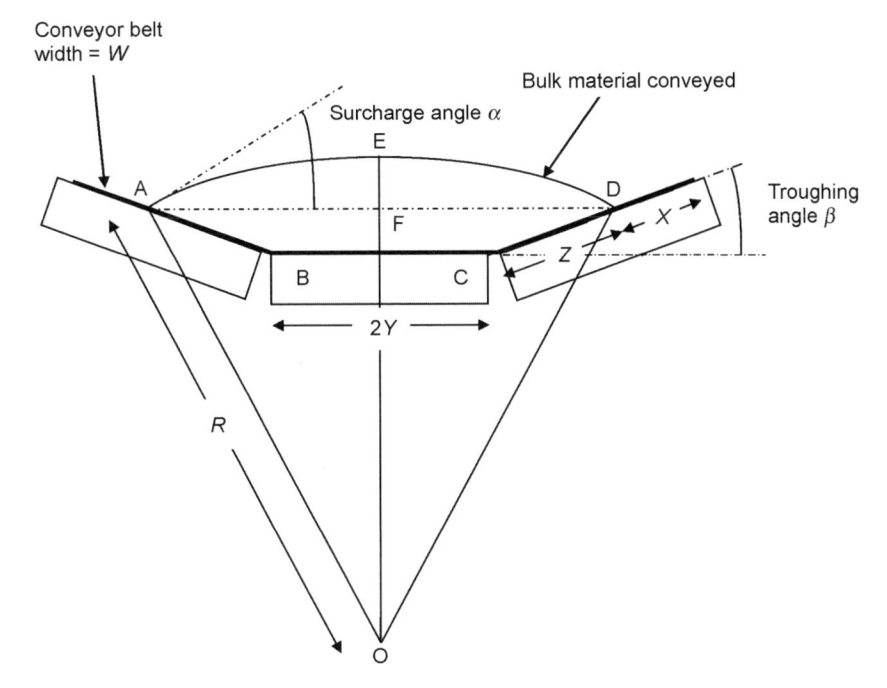

A1.1 Cross-section through troughed conveyor.

$$\Pi.R^2 . \frac{2.\alpha}{2\Pi} = R^2.\alpha$$

when α is in radians, or

$$\frac{\Pi.\alpha}{180} R^2$$

when α is expressed in degrees.

The areas of each of the two triangles OAF and ODF are

$$\tfrac{1}{2}.R.\sin\alpha..R\cos\alpha = R^2.\sin\alpha.\cos\alpha$$

Hence the area *A2* of the segment ADE is

$$R^2 . \frac{\Pi\alpha}{180} - R^2.\cos\alpha.\sin\alpha = \tfrac{1}{2}.R^2 \left(\frac{2\Pi.\alpha}{180} - \sin 2\alpha \right)$$

Substituting for *R*

$$A2 = \tfrac{1}{2} \left(\frac{Y + Z.\cos\beta}{\sin\alpha} \right)^2 \left(\frac{2.\Pi.\alpha}{180} - \sin 2\alpha \right)$$

and the total area of the load is

$$Z.\sin\beta(2..Y + Z.\cos\beta) + \tfrac{1}{2}\left(\frac{Y + Z.\cos\beta}{\sin\alpha}\right)^2 \cdot \left(\frac{2\Pi\alpha}{180} - \sin 2\alpha\right)$$

The load per metre run is then $= \dfrac{\text{cross-sectional area} \times \text{density}}{10^3}$,

where density is measured in kg/m^3. Tonnes per hour is given by tonnes per metre run × belt speed × 3600, when belt speed is measured in m/s. Similar calculations can be made for other idler configurations, such as two-roller or five-roller sets.

Appendix 2
Listing of international, European and national standards relating to belt conveyors

The following is a listing of the principal international and national standards relating to belt conveyors. It is not claimed to be exhaustive.

International Organisation for Standardisation (ISO)

ISO 10247:1990 Conveyor belts – Characteristics of covers – Classification
ISO 1120:2002 Conveyor belts – Determination of strength of mechanical fastenings – Static test method
ISO 14890:2003 Conveyor belts – Specification for rubber or plastics covered conveyor belts of textile construction for general use
ISO 15236-1:2005 Steel cord conveyor belts – Part 1: Design, dimensions and mechanical requirements for conveyor belts for general use
ISO 15236-2:2004 Steel cord conveyor belts – Part 2: Preferred belt types
ISO 15236-4:2004 Steel cord conveyor belts – Part 4: Vulcanised belt joints
ISO 1535:1975 Continuous mechanical handling equipment for loose bulk materials – Troughed belt conveyors (other than portable conveyors) – Belts
ISO 1536:1975 Continuous mechanical handling equipment for loose bulk materials – Troughed belt conveyors (other than portable conveyors) – Belt pulleys
ISO 1537:1975 Continuous mechanical handling equipment for loose bulk materials – Troughed belt conveyors (other than portable conveyors) – Idlers
ISO16851:2004 Textile conveyor belts – Determination of the net length of an endless (spliced) conveyor belt
ISO 1816:1975 Continuous mechanical handling equipment for loose bulk materials and unit loads – Belt conveyors – Basic characteristics of motorised driving pulleys
ISO 18573:2003 Conveyor belts – Test atmospheres and conditioning periods
ISO 2109:1975 Continuous mechanical handling equipment – Light duty belt conveyors for loose bulk materials

ISO 251:2003 Conveyor belts with textile carcass – Widths and lengths
ISO 252:1999 Textile conveyor belts – Adhesive strength between constitutive elements – Part 1: Methods of test
ISO 282:1999 Conveyor belts – Sampling
ISO 283-1:2000 Textile conveyor belts – Full thickness tensile testing – Part 1: Determination of tensile strength, elongation at break and elongation at the reference load
ISO 284:2003 Conveyor belts – Electrical conductivity – Specification and test method
ISO 340:2004 Conveyor belts – Laboratory scale flammability characteristics – Requirements and test method
ISO 3684:1990 Conveyor belts – Determination of minimum pulley diameters
ISO 3870:1976 Conveyor belts (fabric carcass), with length between pulley centres up to 300 m, for loose bulk materials – Adjustment of take-up device
ISO 4123:1979 Belt conveyors – Impact rings for carrying idlers and discs for return idlers – Main dimensions
ISO 433:1991 Conveyor belts – Marking
ISO 5048:1989 Continuous mechanical handling equipment – Belt conveyors with carrying idlers – Calculation of operating power and tensile forces
ISO 505:1999 Conveyor belts – Method for the determination of the tear propagation resistance of textile conveyor belts
ISO 5284:1986 Conveyor belts – List of equivalent terms
ISO 5285:2004 Conveyor belts – Guidelines for storage and handling
ISO 5293:2004 Conveyor belts – Determination of minimum transition distance on three idler rollers
ISO 583-1:1999 Conveyor belts with a textile carcass – Total thickness and thickness of elements – Part 1: Methods of test
ISO 583:1990 Conveyor belts with a textile carcass – Tolerances on total thickness and thickness of covers – Direct measurement method
ISO 703-1:1999 Conveyor belts – Transverse flexibility and troughability – Part 1: Test method
ISO 703:1988 Conveyor belts – Troughability – Characteristics of transverse flexibility and test method
ISO 7149:1982 Continuous handling equipment – Safety code – Special rules
ISO 7590:2001 Steel cord conveyor belts – Methods for the determination of total thickness and cover thickness
ISO 7622-1:1984 Steel cord conveyor belts – Longitudinal traction test – Part 1: Measurement of elongation
ISO 7622-2:1984 Steel cord conveyor belts – Longitudinal traction test – Part 2: Measurement of tensile strength
ISO 7623:1996 Steel cord conveyor belts – Cord-to-coating bond test – Initial test and after thermal treatment

ISO 8094:1984 Steel cord conveyor belts – Adhesion strength test of the cover to the core layer
ISO 9856:2003 Conveyor belts – Determination of elastic and permanent elongation and calculation of elastic modulus
ISO/FDIS 252 Conveyor belts – Adhesion between constitutive elements – Test methods
ISO/FDIS 283 Textile conveyor belts – Full thickness tensile strength, elongation at break and elongation at the reference load – Test method
ISO/FDIS 583 Conveyor belts with a textile carcass – Total belt thickness and thickness of constitutive elements – Test methods
ISO/FDIS 703 Conveyor belts – Transverse flexibility (troughability) – Test method
ISO/TR 5045:1979 Continuous mechanical handling equipment – Safety code for belt conveyors – Examples for guarding of nip points
ISO/TR 8435:1984 Continuous mechanical handling equipment – Safety code for belt conveyors – Examples for protection of pinch points on idlers

European Committee for Standardisation (CEN)

EN 12881-1:2006 Conveyor belts – Fire simulation flammability testing – Part 1: Propane burner tests
EN 12881-2:2005 Conveyor belts – Fire simulation flammability testing – Part 2: Large-scale fire test
EN 12882:2002 Conveyor Belts for General Purpose Use – Electrical and Flammability Safety Requirements
EN 13827:2004 Steel cord conveyor belts – Determination of the lateral and vertical displacement of steel cords
prEN 14973:2004 Conveyor belts for use in underground installations – Electrical and flammability safety requirements
EN 1554:1999 Conveyor Belts – Drum Friction Testing
EN 28094:1994 Steel Cord Conveyor Belts – Adhesion Strength Test of the Cover to the Core Layer (ISO 8094:1984)
EN 618:2002 Continuous Handling Equipment and Systems – Safety and EMC Requirements for Equipment for Mechanical Handling of Bulk Materials Except Fixed Belt Conveyors
EN 620:2002 Continuous handling equipment and systems – Safety and EMC requirements for fixed belt conveyors for bulk materials
EN ISO 1120:2002 Conveyor Belts – Determination of Strength of Mechanical Fastenings – Static Test Method
EN ISO 14890:2003 Conveyor belts Specification for rubber or plastics covered conveyor belts of textile construction for general use (ISO 14890:2003)
EN ISO 15236-1:2005 Steel cord conveyor belts – Part 1: Design, dimensions and mechanical requirements for conveyor belts for general use (ISO 15236-1:2005)

EN ISO 15236-2:2004 Steel cord conveyor belts – Part 2: Preferred belt types (ISO 15236-2:2004)
prEN ISO 15236-3:2005 Steel cord conveyor belts – Part 3: Special safety requirements for belts for use in underground installations (ISO/DIS 15236-3:2005)
EN ISO 15236-4:2004 Steel cord conveyor belts – Part 4: Vulcanised belt joints (ISO 15236-4:2004)
EN ISO 16851:2004 Textile conveyor belts – Determination of the net length of an endless (spliced) conveyor belt (ISO 16851:2004)
EN ISO 20340:1993 Conveyor belts – Laboratory scale flammability characteristics – Requirements and test method (ISO 340:2004)
prEN ISO 22721:2006 Conveyor belts – Specification for rubber – or plastics covered conveyor belts of textile construction for underground mining (ISO/DIS 22721:2005)
prEN ISO 252:2005 Conveyor belts – Ply adhesion between constitutive elements – Test methods (ISO/DIS 252:2005)
EN ISO 252-1:2005 Textile Conveyor Belts – Adhesive Strength between Constitutive Elements – Part 1: Methods of Test (ISO 252-1:1999)
prEN ISO 283:2005 Textile conveyor belts – Full thickness tensile strength, elongation at break and elongation at the reference load – Methods of test (ISO/DIS 283:2005)
EN ISO 283-1 Textile Conveyor Belts – Full Thickness Tensile Testing – Part 1: Determination of Tensile Strength, Elongation at Break and Elongation at the Reference Load (ISO 283-1:2000)
EN ISO 20284:1993 Conveyor belts – Electrical conductivity – Specification and test method (ISO 284:2003)
EN ISO 505:1999 Conveyor Belts – Method for the Determination of the Tear Propagation Resistance of Textile Conveyor Belts
prEN ISO 583:2005 Conveyor belts with a textile carcass – Total belt thickness and thickness of elements – Test method (ISO/DIS 583:2005)
EN ISO 583-1:1999 Conveyor Belts with a Textile Carcass – Total Thickness and Thickness of Elements – Part 1: Methods of Test (ISO 583-1:1999)
prEN ISO 703:2005 Conveyor belts – Transverse flexibility (troughability) – Test method (ISO/DIS 703:2007)
EN ISO 703-1:1999 Conveyor Belts – Transverse Flexibility and Troughability – Part 1: Test Method (ISO 703-1:1999)
EN ISO 7590:2001 Steel cord conveyor belts – Methods for the determination of total thickness and cover thickness (ISO 7590:2001)
EN ISO 7622-1:1995 Steel cord conveyor belts – Longitudinal traction test – Part 1: Measurement of elongation (ISO 7622-1:1984)
EN ISO 7622-2:1995 Steel cord conveyor belts – Longitudinal traction test – Part 2: Measurement of tensile strength (ISO 7622-2:1984)

EN ISO 7623:1997 Steel Cord Conveyor Belts – Cord-To-Coating Bond Test – Initial Test and Test After Thermal Treatment (ISO7623:1996)

EN ISO 9856:2003 Conveyor belts – Determination of elastic and permanent elongation and calculation of elastic modulus (ISO 9856:2003)

Canada (Canadian Standards Association)

CAN/CSA-M422-M87 (R2000) Fire-Performance and Antistatic Requirements for Conveyor Belting

Australian (Standards Australia)

AS 1332-2000 Conveyor belting – Textile reinforced

AS 1333-1994 Conveyor belting of elastomeric and steel cord construction

AS 1333-1994/Amdt 1-2000 Conveyor belting of elastomeric and steel cord construction

AS 1334.1-1982 Methods of testing conveyor and elevator belting – Determination of length of endless belting

AS 1334.10-1994 Methods of testing conveyor and elevator belting – Determination of ignitability and flame propagation characteristics of conveyor belting

AS 1334.11-1988 Methods of testing conveyor and elevator belting – Determination of ignitability and maximum surface temperature of belting subjected to friction

AS 1334.11-1988/Amdt 1-1989 Methods of testing conveyor and elevator belting – Determination of ignitability and maximum surface temperature of belting subjected to friction

AS 1334.12-1996 Methods of testing conveyor and elevator belting – Determination of combustion propagation characteristics of conveyor belting

AS 1334.2-1982 Methods of testing conveyor and elevator belting – Determination of thickness of belting and rubber covers across the width

AS 1334.2A-1984 Methods of testing conveyor and elevator belting – Determination of thickness of cover using an optical magnifier

AS 1334.3-1982 Methods of testing conveyor and elevator belting – Determination of full thickness tensile strength and elongation of conveyor belting

AS 1334.4-1982 Methods of testing conveyor and elevator belting – Determination of troughability of conveyor belting

AS 1334.7-1982 Methods of testing conveyor and elevator belting – Determination of ply adhesion of conveyor belting

AS 1334.8-1982 Methods of testing conveyor and elevator belting – Determination of resistance to tear propagation and resistance of carcass to tearing
AS 1334.9-1982 Methods of testing conveyor and elevator belting – Determination of electrical resistance of conveyor belting
AS 1755 – 2000 Conveyors – Safety requirements
AS 3552-1988 Conveyor belting – Guide to splicing steel cord belting
AS 4035-1992 Conveyor and elevator belting – Glossary of terms
AS 4076.1-1992 Conveyor belts – Determination of strength of mechanical fastenings – Static test method
AS 4606-2000 Fire resistant and antistatic requirements for conveyor belting used in underground coal mines

China

GB/T 10822-2003 Flame retardant conveyor belts of textile construction for general use
GB/T 12736-1991 Conveyor belts – Determination of strength of mechanical fastening – Static test method
GB/T 14784-1993 Safety regulations of belt conveyor
GB/T 15902-1995 Conveyor belts – Determination of elastic modulus
GB/T 16412-1996 Test method for burn test of conveyor belt using propane burner
GB/T 17044-1997 Steel cord conveyor belts – Adhesion strength test of the cover to the core layer
GB/T 17119-1997 Continuous mechanical handling equipment – Belt conveyors with carrying idlers – Calculation of operation power and tensile forces
GB/T 3685-1996 Conveyor belts – Flame retardation – Specifications and test method
GB/T 3690-1994 Conveyor belts of textile carcase – Test method for tensile strength and elongation
GB/T 4490-1994 Conveyor belts – Dimensions
GB/T 5753-1994 Steel cord conveyor belts – Cover thickness measurement
GB/T 5756-1986 Terms of conveyor belts and transmission belts
GB/T 7983-1987 Conveyor belts – Troughability – Method of test
GB/T 7985-1987 Conveyor belts – Carcass tear resistance – Method of test
GB/T 7986-1997 Conveyor belts – Drum friction – Method of test

Germany (Deutsches Institut für Normung)

DIN 15207-1:2000 Continuous mechanical handling equipment – Idlers for belt conveyors handling loose bulk materials – Main dimensions

Appendix 2   177

DIN 15207-2 Continuous mechanical handling equipment; idlers for belt conveyors for unit loads; main dimensions of idlers

DIN 15209 Continuous mechanical handling equipment; belt conveyors, impact rings for carrying idlers

DIN 15210 Continuous mechanical handling equipment; belt conveyors, discs for return idlers

DIN 22100-1 Synthetic materials for use in underground mines; textile-reinforced conveyor belts; safety requirements; testing; marking

DIN 22100-2 Synthetic materials for use in underground mines; steel-reinforced conveyor belts; safety requirements, testing, marking

DIN 22101 Continuous conveyors – Belt conveyors for loose bulk materials – Basis for calculation and dimensioning

DIN 22102-1 Conveyor belts with textile plies for bulk goods; dimensions, specifications, marking

DIN 22102-2 Conveyor belts with textile plies for bulk goods; testing

DIN 22102-3 Conveyor belts with textile plies for bulk goods; permanent joints

DIN 22103 Flame resistant steel cord conveyor belts; requirements and method of test

DIN 22107 Continuous mechanical handling equipment; idler sets for belt conveyors for loose bulk materials; principal dimensions

DIN 22109-1 Conveyor belts with textile plies for coalmining – Part 1: Monoply belts for underground applications; dimensions, requirements

DIN 22109-2 Conveyor belts with textile plies for coalmining – Part 2: Rubber-belts with two plies for underground applications; dimensions, requirements

DIN 22109-4 Conveyor belts with textile plies for coalmining – Part 4: Rubber-belts with two plies for above ground applications; dimensions, requirements

DIN 22109-5 Conveyor belts with textile plies for coalmining; branding

DIN 22109-6 Conveyor belts with textile plies for coalmining – Part 6: Testing

DIN 22110-2 Testing methods for conveyor belt joints – Part 2: Endurance running tests, determination of running time of belt joints at conveyor belts with textile plies

DIN 22110-3 Testing methods for conveyor belt joints; determination of time strength for conveyor belt joints (dynamical testing method)

DIN 22111 Belt conveyors for underground coal mining – Light construction

DIN 22111 Berichtigung 1 Corrigenda to DIN 22111:2000-03

DIN 22112-2 Belt conveyors for underground coalmining – Idlers – Part 2: Requirements

DIN 22112-3 Belt conveyors for underground coalmining – Idlers – Part 3: Testing

DIN 22114 Belt conveyors for underground coalmining – Heavy construction

DIN 22114 Berichtigung 1 Corrigenda to DIN 22114:2001-12
DIN 22115 Belt conveyors for underground coalmining – Coating on idlers – Requirements and testing
DIN 22116 Belt conveyors for underground coal mining – Pressure rollers DN 159 – Dimensions, requirements, marking
DIN 22117 Conveyor belts for coalmining; determination of the oxygen index
DIN 22118 Conveyor belts with textile plies for use in coal mining; fire testing
DIN 22120 Elastomeric scraper plates for belt conveyors in hard coal mines
DIN 22121 Conveyor belts with textile plies for coalmining – Permanent joints for belts with one or two plies; dimensions, requirements, marking
DIN 22122 Continuous mechanical handling equipment – Troughability of conveyor belts – Determination of that portion of width of conveyor belting in contact with idlers; requirements, testing
DIN 22129-1 Steel cord conveyor belts for underground coalmining; dimensions, requirements
DIN 22129-2 Steel cord conveyor belts for underground coalmining; marking
DIN 22129-3 Steel cord conveyor belts for underground coalmining; testing
DIN 22129-4 Steel cord conveyor belts for use in underground coal mining; belt joints; dimensions, requirements
DIN 22131-2 Steel cord conveyor belts for hoisting and conveying; marking
DIN 22131-3 Steel cord conveyor belts for hoisting and conveying; testing
DIN 22434-4 Underground lighting in hard coal mining – Lighting – technological planning bases – Part 4: Conveyor belt
DIN 8132 Continuous totalising automatic weighing instruments (Belt weighers) – Metrological and technical requirements, tests (OIML R 50-1:1997)

Japan (Japanese Standards Association)

JIS B 8803:1990 Rollers for belt conveyor
JIS B 8805:1992 Rubber belt conveyors with carrying idlers – Calculation of operating power and tensile forces
JIS B 8814:1992 Pulleys for belt conveyors

Korea (Korean Standards Association)

B 6182 Portable Belt Conveyors
B 6229 Rollers for Belt Conveyor
B 6279 Pulleys for Belt Conveyor
B 6406 Methods of Calculations and Tests for Rubber Belt Conveyors
M 6534 Ply Construction Conveyor Belts
M 6575 Rubber Belts for Portable Conveyors
M 6639 Measuring Methods for Strength of Mechanical Fastenings for Conveyor Belts

M 6678 Qualitative Standard for Flame Resistance of Conveyor Belts
M 6692 Testing Methods of Steel Cord Conveyor Belts

Sweden

SS-EN 1554 Conveyor belts – Drum friction testing

South Africa

SANS 10266:1995 The Safe use, operation and inspection of man-riding belt conveyors in mines

USA

ANSI/ASME B20.1 Safety Standards for Conveyors and Related Equipment.
ANSI/CEMA 402-2003 Belt Conveyors
ANSI/CEMA 403-2003 Belt Driven Live Roller Conveyors
JJ-B-191A Belting, flat, conveyor or power transmission, cotton (solid woven)

United Kingdom (British Standards Institution)

BS 490-10.1:1983 Conveyor and elevator belting. Testing for physical properties. Introduction
BS 490-2:1975 Conveyor and elevator belting. Rubber and plastics belting of textile construction for use on bucket elevators.
BS 490-10.5:1984 Conveyor and elevator belting. Testing for physical properties. Method for determination of tensile strength and elongation at break of rubber covers.
BS 490-10.7:1984 Conveyor and elevator belting. Testing for physical properties. Method for determination of length of an endless belt
BS 3289:2005 Textile carcase conveyor belting for use in underground mines (including fire performance). Specification.
BS 4531:1986 Specification for portable and mobile troughed belt conveyors
BS 6593:1985 Code of practice for on-site non-mechanical jointing of plied textile and steel reinforced conveyor belting
BS 7300:1990 Code of Practice for Safeguarding of hazard points on troughed belt conveyors
BS 8407:2002 Specification for mechanical and spliced joints in conveyor belting for use underground

Appendix 3

Man-riding conveyors – précis of conditions set out in British Coal Corporation 'Codes and Rules Underground belt conveyors' CR/13

A3.1 Introduction

- Risk of injury to persons using a man-riding conveyor installation are reduced on one hand by good design – on the other by high levels of self-discipline and awareness of potential hazards.
- No conveyor that delivers into a bunker or crusher shall be used for man-riding.

A3.2 Clearances

	Min. clearance	Distance	Older installations
Top belt			
At boarding platform	1.5 m	Belt travel in 4 s beyond platform	
At alighting platform	1.8 m where practicable, Min. 1.5 m	Before leading edge of platform and beyond safety device, belt travel in 1.5 s	Before leading edge of platform and beyond safety device, 3 m each
Elsewhere on the conveyor	900 mm	Along conveyor and past safety device of stopping distance of conveyor plus 3 m	750 mm
Bottom belt			
At boarding platform	1.4 m	Belt travel in 4 s beyond platform	
At alighting platform	1.8 m where practicable, Min. 1.4 m	Before leading edge of platform and beyond safety device, belt travel in 1.5 s	
Elsewhere on the conveyor	900 mm	Along conveyor and past safety device of stopping distance of conveyor plus 3 m	

A3.3 Key dimensions

- Max man-riding speed = 3 m/s (exemptions allowed)
- Max gradient = 1 in 5 (1 in 4 if no mineral conveyed, and max speed 2.67 m/s)
- Max gradient at alighting platform = 1 in 8
- Min belt width = 750 mm with wear no greater than 10%.

A3.4 Men and minerals

- When conveyed together – men and mineral remain 5 m apart
- No haulage to operate alongside belt at main man-riding times or anytime when gradient of roadway exceeds 1 in 10.

A3.5 Inspection roles

- Inspections max 2 hours before man-riding. In man-less operations, inspection patrols at beginning of shift for full length of conveyor and once during shift over man-riding length
- Daily and weekly checks of safety systems and clearances.

A3.6 Boarding and alighting stations

- Whitewashed and lit over length of station plus 9 m at each end
- Signs in front of alighting station at 45 m, 30 m, 15 m and platform
- Suitable notices specifying the following:
    Only board and alight at designated stations
    Face direction of travel
    Remain within confines of belt
    4.5 m spacing between people
    No walking on conveyor
    Wear cap lamps on helmet
    Safety device indicator showing safety device engaged.
- Platform lengths
    Boarding = 1.5 m max.
    Alighting = 6 × belt speed, or 9 m, but not greater than 15 m
- Platform widths
    Boarding = 1 m
    Alighting = 450–1000 mm
- Platform position relative to end of conveyor
    Alighting – stopping distance of conveyor plus 9 m, or 18 m whichever is greater.

A3.7 Safety devices

- Safety device located 3 m max. beyond alighting platform (if at side of conveyor).
- Bottom belt man-riders have plough beyond safety device at stopping distance plus 3 m.
- Safety device will cut off power and apply brakes if activated.
- Slip detector device fitted.
- Belt alignment device fitted.
- Belt breakage, tear detection and chute blockage devices fitted.
- Arrangement to prevent inadvertent reversal of belt.

Index

Abra copper mine, Chile 166
acceleration factor 41–2
accelerator conveyor 65
acceptance criteria in fire propagation testing 120, 124, 128, 129–30
adhesion tests 100–1, 105
alighting stations 149, 181
angle of wrap 30–1, 34
anti-runback device 59
anti-seize bearings 50
Aracoma Coal Company fire 136, 137
artificial friction coefficient 23, 24
ATH Resources 162–5
Australia
 standards 175–6
 studies on fire safety 128–32

ball bearings 50
Barclay's fire propagation test 119–21
Barthel burner 120
basic CEMA method 26
Batu Hijau gold mine, Indonesia 166–7
bearing life 48–9, 50
belt capacity 21, 168–70
belt cleaners 66–8, 155
belt construction 71–83
 cable belts 71, 82–3, 159
 designations 78, 81
 and joint strength 87–8
 steel cord belts *see* steel cord belts
 textile carcase belts *see* textile carcase belts
belt failure 35–8, 160–1
belt hangers 147, 148
belt length 152

belt sag 38, 49
belt slip *see* slip
belt speed *see* speed, belt
belt strengths 77, 79–80
belt tension *see* tension
belt tracking (training) 153–4, 155
belt weigher 57–8
belt width 19–20
bend pulleys 17, 18, 53
bending stress 35
'biggest and best' conveyor installations 165–7
blistering 36, 76–7
blockages 63, 146
 from fines deposits 64
 from large lumps/objects 63–4
boarding stations 149, 181
booster drives 42, 58–9
brakes 59
'branding' 78, 81
Brandstrecke test 126, 127, 128
break, elongation at 99–100
'breakers' in belts 73, 79, 81, 91–2
'British Coal' approach to power requirements 24–6, 28–9
British Coal Corporation Codes and Rules 148, 180–2
British Standards Institution (BSI) 18, 96–7, 179
 BS 8438 22
 tests of joint strength 105–6
British Steel Corporation 97
bunkers 8–9, 149–50
Bunsen burner 119, 120
butt spliced joints 90

Index

cable belts 71, 82–3, 159
Canadian Standards Association 97, 175
capacity, belt 21, 168–70
carry side tension 30, 31–3
carrying idlers 17, 18
case histories 158–67
   ATH Resources 162–5
   'biggest and best' 165–7
   Creswell Colliery fire 72, 110–12
   low stretch textile carcase belt 35–8, 160
   Prosper Haniel mine 161–2
   Selby mine 65, 156, 158–61
CEN *see* European Committee for Standardisation
centralisation, flow 65–6
centre idler 17, 18, 44
chamfering 87
chemicals, damage due to 153
Chhatak cement plant–Lafarge quarry conveyor 166
China 176
chute design 63–6
cleaners, belt 66–8, 155
clearances 155, 180
Cliffe Hill granite quarries 12
clip fasteners 84–5
coal mines 1–2, 5–9
   belt conveyors at the coalface 5–7
   fire hazards 109–10
   trunk conveyors 7–9
   see also under individual mines
coefficient of friction 30–1, 34, 54, 55–6
   artificial 23, 24
complex drives 31–5, 55–6
concave curves 41–2
concentricity of idler barrels 50
cone calorimeter test 131, 132
constant tension winch systems 61
continuous dynamic pulsating stress testing 107
convex curves 41–2
Conveyor Equipment Manufacturers' Association (CEMA) 10, 18, 19
   approach to power requirements 26–9
   basic method 26
   standard method 26–7
   universal method 27–8
conveyor structure 46–7, 51–2, 154

cord-to-coating bond test 104
cores 152
costs
   comparison of forms of haulage 9–10, 15
   truck haulage 11
cotton duck 4, 5
covers 75–7, 81, 153
   testing properties of 105
   testing thickness of 100, 103
creep 34–5
Creswell Colliery fire 72, 110–12
cross-sectional area 21, 168–70
crowned pulleys 53, 154
crushers 149–50
   in-pit 11, 12–13

deep groove ball bearings 50
Deep Mined Coal Advisory Committee 110
delamination 36, 72
delays *see* stalled belts
Dendrobium mine, New South Wales 9
design 17–70
   basic considerations 19–21
   belt capacity 21, 168–70
   belt cleaning 66–8
   belt tensions 29–39
   conveyor idlers 47–50
   conveyor structure 51–2
   drives 54–9
   high angle conveyors 68–9
   power requirements 21–9
   pulleys 52–4
   tension changes over the belt width 39–42
   tensioning devices 60–2
   transfer points 62–6
Deutsches Institut für Normung (DIN) 18, 176–8
dynamic testing of joints 106–7
gallery 133
double burner (DB) test 126, 127, 128
drive factor 30–1
drive pulleys 17, 18, 53
   distance between 35
drives
   complex 31–5, 55–6
   design 54–9

drum friction tests 114–15, 116–18, 124, 135
dual-pulley dual-motor drive 32–3
dual-pulley geared tandem drive 31–2
dual-pulley ungeared primary plus geared tandem secondary drive 33–4, 35–8
dump trucks 9–11, 14–15
dust explosions 137
dust suppression sprays 63, 64
dynamic strength of joints 105–7

elastic effects 34–8
elastic modulus 101
elastic stretch 60, 101
electrical resistance tests 138–9, 140
electrostatic hazards 137–40
elongation (stretch)
  at break 99–100
  elastic and permanent 60, 101
  low stretch belts 35–8, 160
  at reference load 99–100
  testing steel cord belts 104
empty belt, power to move 24, 25
enclosed chutes 63, 64
environmental concerns 136
Essential Health and Safety Requirements (EHSR) 97
European Committee for Standardisation (CEN) 96, 97–8
  fire safety standards 134–5
  specific standards and tests 98–105, 173–5
European Community/Union 97
  Mines Safety and Health Commission 121–5

face seals 48
factor C 23, 25
factor of safety (service factor) 39, 77–8, 80, 88
Factory Mutual Research Corporation (FMRC) test 131, 132
feeder-breaker units, mobile 65
fines deposits 64
finger spliced joints 90–2
fire propagation tests 115–21, 122–3, 124–8, 129–30, 132
fire safety 96–7, 109–37

Australian studies 128–32
Barclay's fire propagation test 119–21
belt construction 72–3
Cresswell Colliery fire 72, 110–12
early research into conveyor fires 112–19
  fire propagation 115–19
  fires caused by friction 114–15, 116–18
European Community 121–5
fire hazards 109–10
idlers 50
large-scale gallery tests 124–8, 129–30, 131–2, 133, 134, 135–6
mid-scale gallery tests 133–5
small-scale tests 119–21, 122–3, 131–2, 135–6
fixed fill fluid coupling 56–7
flame propagation index (FPI) 132
flame spread 131
flame test 115, 119
flat helix turnover 68
flow centralisation 65–6
fluid couplings 33, 56–7
Fosbucraa conveyor system 166
four motor, four drive pulley ungeared drive 33–4
friction
  coefficient of *see* coefficient of friction
  fires caused by 110–11, 112, 113
    drum friction tests 114–15, 116–18, 124, 135
full thickness tensile testing 99–100

gallery tests 121–8
  large-scale 124–8, 129–30, 131–2, 133, 134, 135–6
  mid-scale 133–5
garland (catenary) idler 45–6
gate conveyors 5–7, 9
General Foods dust explosion 137
general machinery monitoring procedures 156–7
German standards *see* Deutsches Institut für Normung (DIN)
granaries 4
granite quarrying 11, 12
gravel and sand operations 13–15
gravity take-up devices 60, 61

gravity towers 60, 61, 62
guarding 150, 163, 164
  of nip points 144–6
guide rollers 154

handling of belts 152–3
head end 6, 17, 18
Health and Safety Executive (HSE) 144, 145, 149
heat, damage by 153
heat release rate (HRR) 132
high angle conveyors 68–9
high density polyethylene (HDPE) 51
high energy test 126, 127, 128, 132
hinged joints 85, 86
history 4–16
  belt conveyors in mines 5–9
    coalface 5–7
    trunk conveyors 7–9
  early applications 4–5
  stone quarries 9–15
    in-pit crushing and screening 12–13
    sand and gravel operations 13–15
    surface mining 13, 14
Horizon mining 7
horizontal curves 42, 59, 163
horizontal movement of load, power for 24, 25
hydraulic winch 60

idlers 4, 17, 18, 44–51
  bearing failure 155
  belt tracking 153–4
  design of 47–50
  idler sets 44–7
  maintenance 155
  non-metallic 51
  spacing 38, 49–50
impact idlers 47, 65
in-line idler sets 44, 45, 46
in-pit crushing 11, 12–13
interlaced stepped joints 92
International Electrotechnical Commission (IEC) 97
International Federation of the National Standardising Associations (ISA) 97
International Standardisation Organisation (ISO) 18, 96, 97, 98

fire propagation test 119–21, 124
ISO 5048 approach to power requirements 22–4, 28–9
specific standards and tests 98–105, 171–3

Japanese Standards Association 178
joints 84–94
  failures 88–9
  mechanical fasteners 84–9
  spliced joints 84, 89–93, 156
  strength 87–8
  testing 102, 103, 105–7
jump spliced joints 90

Korean Standards Association 178

labyrinth seals 48
Lafarge quarry–Chhatak cement plant conveyor 166
lagging 54
large lumps/objects 6, 63–4
large-scale gallery tests 124–8, 129–30, 131–2, 133, 134, 135–6
length, belt 152
lifting equipment 153
lignite surface mines 13, 14
limiting slip 30, 34
loading points *see* transfer points
locomotive haulage 7, 9–10
Longannet mine 82
longitudinal traction tests 104
longwall coal mining 10
loop take-up devices 62
low stretch textile carcase belt 35–8, 160
lump-breakers 149–50
lump size
  and belt width 19–20
  sizers 8, 10, 64, 149
Luxembourg test (two metre test) 124–6, 127, 128, 132

main resistances 22, 23
maintenance 152–7
  belt tracking 153–4
  monitoring belt condition 155–7
  optimising belt life 154–5
  supply, storage and handling of belts 152–3

man-riding
  British Coal Corporation Codes and
    Rules 148, 180–2
  joint failures 88–9
  safety 146–50, 182
manager's transport rules 150
manpower requirements 8, 63
materials characteristics 19
materials handling safety 150
mean stripping force 100
mechanical fasteners 84–9
  testing strength of 102, 103
Mersey Docks and Harbour Board 4
mid-scale gallery tests 133–5
mildew attack 76, 153
Mine Safety and Health Administration
  (MSHA) test 133–4
Mines Safety and Health Commission
  121–5
Mini Maritsa Iztok (MMI) 13
minimum tension 38–9, 49
mobile feeder-breaker units 65
monitoring belt condition 155–7
motorised pulleys 58
multiple motor drives 56
multi-pulley drives 31–8, 55–6

National Coal Board (NCB) 96–7, 112,
  120, 121
nip points 142–6
non-destructive testing 156
non-metallic idlers 51
nylon 51

offset (staggered) idler sets 44, 45, 46,
  155
oils 153
omega drive configuration 35–8, 56
optimisation of belt life 154–5
ozone 153

Pelambres copper mine, Chile 166
permanent stretch 60, 101
Phoenix Conveyor Belts 162, 166
piggy-back boosters 59
piles (protective surface yarns) 73, 74
plain stepped joints 92, 93
plate joints 85, 86–7
plied belts 4, 5, 55, 71, 72, 74, 76

spliced joints 89–90
polyamide 73–4
polyester 73–4
polymeric idlers 51
polyurethane 91
polyvinyl chloride (PVC)
  belts impregnated with 71, 75–6, 90
  bonding in spliced joints 91–2
  covers 75–6
  fire resistance 72–3, 113
power requirements 21–9
  'British Coal' approach 24–6, 28–9
  CEMA approach 26–9
  comparison of power calculations 28–9
  ISO 5048 approach 22–4, 28–9
propane burner flame tests 124–8, 129–30
Prosper-Haniel mine 161–2
protective surface yarns (piles) 73, 74
pulleys
  belt tracking 153–4
  design 52–4
  distance between 35

quarries 9–15
  in-pit crushing and screening 12–13
  lignite surface mines 13, 14
  sand and gravel operations 13–15

radius of curvature 41
rail haulage 7, 9–10
reference load, elongation at 99–100
resistance
  electrical 138–9, 140
  to motion 22–4
return end (tail end) 6, 17, 18
return idlers 17, 18, 45
rips 79
  detection 156
risk assessment 135, 145
Rotacure process 76
rotational torque of idlers 48
rubber 4, 5
  covers 76–7, 81
  protection from attack 153
RWE Power AG 13

safety 2, 109–51
  electrostatic hazards 137–40
  fire safety *see* fire safety

man-riding 146–50, 182
materials handling 150
nip point accidents 142–6
stored energy 146
sag, belt 38, 49
sand and gravel operations 13–15
scab spliced joints 90
scraper cleaners 66–8
screening, in-pit 12–13
screw-type take-up devices 60
seals 47–8, 50
secondary resistances 22, 23
secondary safety devices 135
Selby mine 65, 156, 158–61
self-aligning idlers 154
service factor (factor of safety) 39, 77–8, 80, 88
sideways movement 52, 53
signage 145, 150
sizers 8, 10, 64, 149
skive spliced joints 90
slack side tension 30
slider beds 65–6
slip 34
  detectors 155
  limiting slip 30, 34
slope resistance 23–4
small-scale fire tests 119–21, 122–3, 131–2, 135–6
snub pulleys 17, 18
'soft start' technology 42, 56–7
solid woven belts 4, 71–6, 77
  finger spliced joints 90–2
South Africa 179
South Spine conveyor, Selby mine 160–1
special main resistances 22, 24
special secondary resistances 22, 24
speed, belt 19, 21, 65
  and man-riding 147, 148, 149
  speed control 57–8
spillages 63, 143, 146
spine roads 158–9
spliced joints 84, 89–93, 156
staggered (offset) idler sets 44, 45, 46, 155
stalled belts 8, 54, 113, 155
standard CEMA method 26–7
standardisation 2
  process 96–8

standards 2, 18, 95–108, 171–9
  Australia 175–6
  Canada 175
  CEN see European Committee for Standardisation
  China 176
  Germany see Deutsches Institut für Normung (DIN)
  ISO see International Standardisation Organisation
  Japan 178
  Korea 178
  South Africa 179
  Sweden 179
  UK see British Standards Institution
  USA 179
Standards Australia 175–6
staple fasteners 85–6
starting torque 56
steel cord belts 35, 40, 78–81, 160–1, 162
  monitoring 156
  spliced joints 92–3
  testing 103–5
step spliced joints 89–90, 92–3
stools 52
storage of belts 152–3
stored energy 146
strength
  belt 77, 79–80
  joints 87–8
  testing 102, 103, 105–7
stresses 35
  optimising belt life 154–5
stretch see elongation
stringers 52
structure, conveyor 46–7, 51–2, 154
supervisory control and data acquisition (SCADA) system 163–4
supply of belts 152–3
surcharge angle 17, 18, 20
surface mines
  ATH Resources 162–5
  coal drift mines 7–8
  lignite mines 13, 14
surface resistance tests 138–9, 140
Sweden 179

tail end (return end) 6, 17, 18

take-up devices 17, 18, 60–2, 154
tandem conveyor 159–60
tandem drives 55, 56
taper roller bearings 50
tear propagation resistance 101–2
tensile strength tests 99–100, 102, 103, 104
tension 29–39, 55, 77
   basic consideration 29–31
   CEMA approach to power requirements 26, 27, 28
   complex drives 31–5
   factor of safety 39, 77–8, 80, 88
   low stretch textile carcase belt 35–8
   minimum tension 38–9, 49
   and stored energy 146
   tension changes over the belt width 39–42
tensioning methods 60–2
testing/tests 95–108
   electrical resistance 138–9, 140
   fire safety 114–36
   ISO 98–105
   joint strength 102, 103, 105–7
   monitoring belt condition 155–6
   standardisation process 96–8
textile carcase belts 71–8
   spliced joints 89–92
   tests 99–102
   see also plied belts; solid woven belts
thermal response parameter (TRP) 132
thickness testing 100, 103
torque control 57–8
torque converter 57
total thickness tests 100, 103
tracking, belt 153–4, 155
training of personnel 143, 144–5
tramp metal 155
transfer points 8, 155
   design 62–6
transition distance 39–41
transverse flexibility 102–3
transverse reinforcement 79, 80, 81

tripper boosters 17, 59
tripper-stacker 163, 165
troughability 102–3
troughing angle 17, 18, 20–1, 155
truck haulage 9–11, 14–15
trunk conveyor belts 7–9
turnovers 68
two drive pulley system 56
two geared tandem drive 34
two metre propane burner test 124–6, 127, 128, 132

ultraviolet (UV) light 152–3
United Kingdom (UK) standards *see* British Standards Institution
United States (USA) standards 179
universal CEMA method 27–8
universal control drive (UCD) 57–8

variable fill couplings 57, 58
variable frequency drives (VFDs) 58, 61
variable speed drives (VSDs) 58
vertical curves 41–2
vertical movement of load, power for 24, 25–6
visco-elastic effects 34–8
vulcanised joints 80, 92–3, 105–6
vulcanising 76

warp threads 73
wear indicator 77
weft (steel reinforcement) 79
weft threads 73
   and joint strength 87–8
wet clutch 56, 57
width, belt 19–20
winch take-up devices 60, 61
wing idlers 17, 18, 44
wire hook fasteners 85–6
wrap, angle of 30–1, 34

X-ray examination 156